Orlicky's Material Requirements Planning

Orlicky's Material Requirements Planning

George Plossl

Second Edition

McGraw-Hill, Inc.

New York San Francisco Washington, D.C. Auckland Bogotá
Caracas Lisbon London Madrid Mexico City Milan
Montreal New Delhi San Juan Singapore
Sydney Tokyo Toronto

Library of Congress Cataloging-in-Publication Data

Orlicky, Joseph.
[Material requirements planning]
Orlicky's Material requirements planning / George Plossl.–2nd ed.
p. cm.
First pub. under title: Materials requirements planning / Joseph Orlicky, New York : McGraw-Hill, 1975.
Includes bibliographical references and index.
ISBN 0-07-050459-8
1. Production control–Data processing. 2. Inventory control–Data processing. 3. Manufacturing resource planning. I. Plossl, George. II. Orlicky, Joseph. Material requirements planning. III. Title.
TS155.8.074 1995
658.7–dc20 93-41951
CIP

The first edition of this book, by Joseph Orlicky, was published by McGraw-Hill in 1975 under the title *Material Requirements Planning.*

10 11 12 13 14 DOC/DOC 0 9 8 7 6 5 4 3 2

ISBN 0-07-050459-8

The sponsoring editor for this book was James H. Bessent, Jr., the editing supervisor was Peter Roberts, and the production supervisor was Donald Schmidt. It was set in Baskerville by Judith N. Olenick.

Printed and bound by R. R. Donnelley & Sons Company.

This book is printed on recycled, acid-free paper containing a minimum of 50% recycled de-inked fiber.

This book is dedicated to the practitioners—managers, supervisors, planners, programmers—and all others who struggled to make MRP work, to the support staff people who sought the reality behind the MRP myths, to students who tried to learn solutions to problems they didn't understand, and to our wives, Olga and Marion, who had to live with the whims and fancies of two authors.

Contents

Figures and Tables

Figures

Tables

Foreword

In 1966, Joe Orlicky, Oliver Wight, and I met in an American Production and Inventory Control Society (APICS) conference. We found that we had all been working on material requirements planning (MRP) programs, Joe at J.I. Case Company and IBM, Oliver and I at The Stanley Works. We continued to meet and compare notes on MRP and other topics. In the early 1970s we organized the APICS MRP Crusade, using the resources of the Society and the knowledge and experiences of a few "Crusaders" to spread the word on MRP among APICS members and others interested. All but a few APICS chapters participated.

This crusade showed clearly the need for more professional literature and teaching aids on this powerful but new technique. Oliver's and my book, *Production and Inventory Control: Principles and Techniques*, published in 1967, was the first to include coverage of MRP—all of 16 pages' worth! Users of MRP were too busy trying to make it work to take time to write about it.

Joe himself said it best in his first-edition Preface:

> I found that the entire MRP literature consisted of 26 items—good, bad, and indifferent—all of which were either articles, excerpts, special reports, or trade-press 'testimonials.' Tutorial material on MRP basics was lacking entirely. The job of "getting it all together" remained to be done. Someone had to write a book on "MRP from A to Z" and I concluded I may have to be the one. So I wrote this book.

We were delighted when Joe began work and several of us offered assistance, mainly suggesting topics to include and critiquing his work. He needed little of either help; the success of his book showed the quality of his work. Since 1975, the first edition has served as the definitive text on MRP. In characteristically careful, deliberate, and thorough style, Joe Orlicky assembled the relevant body of knowledge and produced his classic work, the first full coverage in the literature.

Almost two decades have passed and much has changed in the field of manufacturing planning and control, some for the better and some for the worse. MRP has become the unchallenged core planning program for most manufacturing companies. Much more powerful and far less expensive computer hardware and a multitude of software programs are now available to bring practical applications of MRP within the grasp of everyone needing them. Many applications show the great potential benefits.

But many have struggled with MRP with only partial success; the reasons are presented in this second edition. Lack of understanding at all organization levels lies behind most of these reasons: too few know how manufacturing works, how it can be planned and controlled best, the roles of MRP, its requirements, and its use. We were only beginning to comprehend these when Joe wrote his book. They are now well understood and I have included them in this second edition.

It was a signal honor to be asked to make this revision, and it has been a challenging task. I have kept the best, discarded only the outdated, and added important interim developments.

George W. Plossl

Preface

The purpose of this book is to provide comprehensive coverage of a technique of first importance to manufacturing operations management—computer-based material requirements planning. MRP, as it is commonly called, represented a sharp break with past theory and traditional practices in both its underlying philosophy and methods. Broadly viewed, the development of MRP programs marked the coming of age of the field of production and inventory control and introduced a new way of life in the management of all manufacturing businesses.

Computer hardware and software became commercially available in the early 1960s and were capable of handling data in volumes and at speeds previously scarcely imaginable; these lifted data processing constraints that had handicapped inventory management. They made obsolete the older methods and techniques devised to live with these limitations.

In the early days, the most significant benefits were not achieved by those pioneering manufacturing firms that chose to improve, refine, and speed up existing procedures with computers but by those who undertook fundamental changes to their planning and control systems. However, abandoning familiar techniques, even those that had proven unsatisfactory, and substituting new, radically different approaches like MRP-based systems required new knowledge not possessed by many people at that time. The first edition of this book provided an authoritative source of information on MRP to help develop the needed knowledge.

Since the early 1960s, when a few of us pioneered the development and installation of computer-based MRP systems, time-phased material requirements planning has come a long way—both as a useful technique and as a source of new knowledge. From the original handful, the number of MRP systems in use in American industry gradually grew to about 150 in 1971, when the growth curve began a steep rise. By 1975 over 700 systems had been implemented, and the 1980s saw thousands more come into use. Today, it is a rare company in Western and Pacific Rim industrial countries that does not operate a material requirements plan. I have deliberately avoided saying that most use them successfully. Too few do! The important reasons are covered in this book.

In the early days, the subject of material requirements planning was neglected in academic curricula, in favor of intellectually challenging statistical and mathematical techniques. People in industry thought these irrelevant or obsolete. Academicians considered the study of MRP "vocational" rather than "scientific." Production and inventory management *are* vocational, of course, in the sense that knowledge is applied to solving real business problems. Like engineering or medicine, production and inventory management are oriented toward practice but must be based on a sound body of theoretical knowledge.

In the past also, communications were poor between practitioners in industry and those who taught production planning and inventory control. Over almost two decades, the first edition of this book helped to improve such communications, because it combined both theoretical knowledge and practical experience.

More recently, material requirements planning has been incorporated into the curricula of many colleges, college extension services, universities, and schools of business. Joe Orlicky's book has been a popular text. The revisions in this second edition will enhance its use in more up-to-date, comprehensive courses. The material on research opportunities will challenge academics to increase their contributions to this body of knowledge in ways that have the highest relevance in the real world.

This second editon of Joe Orlicky's *Material Requirements Planning* will interest three different audiences. Manufacturing executives and managers will find nontechnical explanations of why they should embrace this "way of life in production and inventory management" and learn their responsibilities. Those interested in technical details will see all of the mechanics of MRP and will obtain an important perspective: the users' point of view, their problems, and the skills they need to use the new tools properly. Users and students will find comprehensive treatment of material requirements planning and good coverage of related activities in manufacturing planning and control.

This book is not meant to serve as a text on the general subject of production and inventory management, or even inventory control. While not a prerequisite, the value of this book will be enhanced by at least an elementary knowledge of physical methods of inventory control and manufacturing operations. The book is written primarily for users and potential users of material requirements planning—manufacturing, materials, and production control managers, inventory planners, systems analysts, and interested industrial engineers. It can also aid instructors and students of manufacturing planning and control in gaining a clearer understanding of how manufacturing works with MRP.

The focus of this book is on material requirements planning as a technique. Its objective is to explain procedural logic, functions, and use of MRP programs. The computer's contribution to MRP is simply its power to store and access masses of data and to execute many logical calculations quickly. Hence, computer processing and programming and other purely technical system implementation topics are excluded; these are amply documented in manuals published by computer manufacturers and software suppliers. A bibliography of books and articles is included for those who want more information on related topics.

Willa Cather said, "The one education which amounts to something is learning how to do something well, whether it is to make a bookcase or write a book." I hope I have done the latter so that readers can use it to do something well for themselves, their businesses, and our country's industrial economy.

George W. Plossl

PART 1

The Whole and Its Parts

The farther back you can look, the farther forward you are likely to see.

SIR WINSTON S. CHURCHILL

1
Manufacturing as a Process

GURU: To manage any manufacturing company, Pilgrim, all you need to know is how to read the numbers.

PILGRIM: Yes, Guru, but which numbers?

What Manufacturing Is, and How It Works

Most economists seem to understand very little about manufacturing. They see its effects—contribution to gross domestic product, inventory and business cycles, total employment changes, and trade—but they handle these statistically, making assumptions about the future based on the past, as if managers could make no significant changes. Political leaders understand it even less; to them it's just a prime source of revenue for government to use to cure social problems, many of which they think are caused by manufacturing. To environmental ideologues, manufacturing is the great polluter and destroyer. To labor unions, it is the enemy to be fought in the name of the natural right of workers to get more reward for less effort. To its own workers it is bread and butter, but to the majority of other people it is greedy business reaping enormous profits. Manufacturing has many more detractors than supporters. Employment, called "jobs," is necessary and good, but "employers" seem to be undesirable and bad!

Many efforts have been made to transplant successful manufacturing planning and control practices from one company to another. Such piecemeal changes more often fail than succeed, particularly when crossing national borders. Grafting strong legs on a weak torso cannot produce an athlete. Manufacturing, like humans, *must be strong and effective in all essential elements* to be healthy. An understanding of how manufacturing really works is the key to competitive success.

By definition, manufacturing is the conversion of low-value materials to higher-value products. These must have value at least equal to their price in meeting customers' needs or desires. People, capital, materials, and other resources are employed at costs that must be less than the selling price, taxes must be paid, and, hopefully, money will be left to fund research and development, expand the business, and reward the owners who provide the operating capital.

The major organization segments of a manufacturing firm—marketing and sales, engineering, finance, and production—are as familiar as the groups involved—customers, suppliers, blue- and white-collar workers, unions, executives, government regulators, and many others, even consultants. Few, however, realize that manufacturing is a process in which all these groups participate and within which each is constantly striving for a larger share as a reward for its services. As in society, this infighting for a larger slice of the pie often takes precedence over cooperative efforts to make the pie larger. Such adversarial relationships and lack of motivation to improve may be the primary causes of the inevitable failure of socialism; they are certainly major problems today in our form of capitalism.

Recognition that manufacturing is a process is essential to understanding how it should work. A bewildering variety and great complexity of products, materials, technology, machines, and people skills obscure the underlying elegance and simplicity of the total process, as shown in Figure 1-1. In words,

The essence of manufacturing is flow of materials from suppliers, through plants to customers, and of information to all parties about what was planned, what has happened, and what should happen next.

This is true regardless of what is made, how and when it is made, and who makes it where. The First Law of Manufacturing is

All benefits will be directly proportional to the speed of flow of materials and information.

This is a universal law applying to every type of manufacturing. Difficulties in controlling manufacturing will decrease and planning will become more effective as material and information flows speed up. The best use of resources comes from eliminating problems that interrupt or slow down these flows. The obvious principle that emerges is

Time is the most precious resource employed in the manufacturing process.

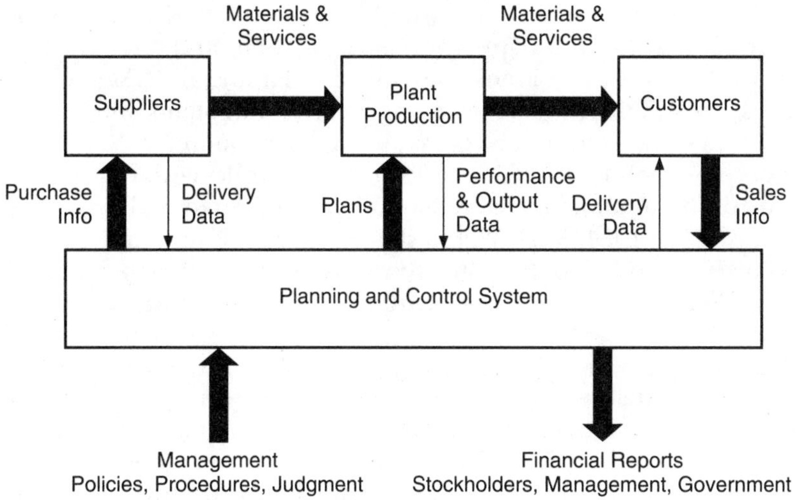

Figure 1-1. The essence of manufacturing.

Equal amounts are available to everyone, but time moves relentlessly; it cannot be stored, extended, or recycled, and wealth can buy no more. Wasting time causes irretrievable loss.

Simple, universal logic underlies all manufacturing and can be represented by six simple questions:

1. What is to be made?
2. How many, and when are they needed?
3. What resources are required to do this?
4. Which are already available?
5. Which others will be available in time?
6. What more will be needed, and when?

Business and marketing strategies determine the answers to the first two questions. Internal company planning and control systems provide answers to the last four.

Evolution of the Art

Manufacturing planning and control theory and practice have been evolving in the United States at an accelerating rate through the last half of the twentieth century. The first rigorous attempts to improve manufacturing processes and workers' productivity were made around the beginning of this century by pioneer industrial engineer, Frederick W. Taylor. The standard-setting techniques developed by him, and later applied widely by Henry Gantt, Frank and Lilian Gilbreth, and Harrington Emerson, are still the basis for planning labor requirements and paying incentives to workers to reward them for producing more.

Ford Harris published in 1915 the first formula to calculate an economic order quantity to minimize the total of ordering-related and inventory carrying costs. In 1934, R. H. Wilson showed how statistics could be used to plan inventory cushions to reduce the impact of forecast errors, reduce material shortages, and improve customer deliveries with minimum inventories.

During WWII, teams of British scientists applied mathematical techniques and scientific methods to complex problems involving the choosing among alternate uses of scarce resources. For over two decades after the war, attempts were made in Europe and America to apply tools of Operations Research, as it was called, mainly linear programming and queuing theory, to industrial logistics. These produced good results in some situations but had only limited application. Too often, operations researchers resembled people with a familiar tool looking for something to fix, like a person with a screwdriver tightening all the screws around and then filing slots in nail heads to tighten these too. Real problems went unattacked and unsolved.

Business computer hardware and software became generally available in the early 1960s, making record keeping and use of complex planning techniques practicable for the myriad of items found even in small manufacturers. This removed obstacles to the development of many planning techniques impractical to apply manually. Prominent among these were:

1. George E. Kimball's "base stock system," aimed at eliminating wide variations in upstream demand caused by independent ordering of components of assembled products. This 1950s system foreshadowed Japanese Kanban "pull" techniques by communicating end-product actual demands to each work center producing components and to outside suppliers. Lack of computers to handle the masses of data, long setups, large component order quantities, and buffer stocks at many process steps prevented wide early use and any reaping of the potential benefits of this technique.

2. A forecasting technique called Exponential Smoothing was publicized by Robert G. Brown in 1959. This weighted averaging technique found wide application in product forecasting because of small computer data storage requirements and flexibility in reacting to demand changes.

Like many other mathematical techniques, variations were developed extending it to unusual demand patterns well beyond the point of diminishing returns.

3. Material requirements planning (MRP) driven by a master production schedule (MPS) was first applied successfully in 1961 by J. A. Orlicky on J. I. Case Company farm machinery. The rigorous logic and masses of data to be handled made these an ideal computer application. The enormous potential benefits over existing ordering techniques generated great interest worldwide.

4. Detailed capacity requirements planning (CRP) was known since Taylor, et al., had showed how to develop work standards. While good computer software was available early in the 1960s, failure to develop adequate capacity planning resulted from poor-quality, incomplete processing data and work standards in most companies. Rough-cut capacity planning techniques were used only for testing the validity of master production schedules. This neglect of capacity requirements planning contributed greatly to the early failures of MRP to realize its full potential.

5. Input/output (I/O) capacity control. Tight control of work input and output was impossible without sound capacity plans. This delayed and blunted attacks on long cycle times and made priority planning and control much less effective.

6. Operation simulation. O. W. Wright, J. D. Harty, and I developed at Stanley Tools in 1962 one of the first detailed computer simulations of plant operations. Our SWIFT (Simulated Work Input and Flow Times) program showed us clearly the harmful effects of high work-in-process levels and erratic work input rates on schedule performance and costs.

Operations researchers were intrigued by the problem of determining "optimum" work sequences in complex manufacturing environments. Sophisticated and expensive computer programs to plan such optimum schedules began to appear in the early 1970s; Werner Kraus of IBM explained his program for Model 301 computers to me over dinner in Stuttgart in mid-1972. As bigger, faster computers became available, IBM people in Great Britain converted KRAUS into CLASS (Capacity Loading And Sequence Scheduling) for the Model 401 computers and later to CAPOSS (CApacity Planning and Operation Sequence Scheduling) for 370 Series and later computers. Other computer software suppliers and independent programmers in Europe and America developed similar optimizing programs, but successful applications were the rare exception.

One of these was R. L. Lankford's 1971 homegrown program at Otis Engineering in Texas; this worked very well for him in predicting schedule falldowns and altering processing to avoid them. The problem with most such "finite loading" was not fatal flaws in the programs but lack of support of sound planning, execution, and control. Fine tuning is useless without a working television set.

Increasing interest in manufacturing planning and control, and a growing need for education, led to the founding of the American Production and Inventory Control Society (APICS) in 1957. Chapters were quickly formed in the United States, Canada, Mexico, Europe, and South Africa, providing a forum to disseminate information among practitioners in industry, consultants, academics, and others interested. The body of knowledge was codified, its language formalized, and examinations developed to test individuals' professional competence by APICS Certification Program Council, which I led in the early 1970s.

Electronic data collection equipment became available in the late 1960s to track activity on plant floors and on receiving and shipping docks, feeding data to core system files directly. This equipment evolved over the next two decades into intelligent work station terminals with enhanced power to improve execution. By the mid-1970s integrated core systems had become a reality in leading companies, although consistent improvement and notable benefits eluded most firms using them in industrial countries.

The problems handicapping systems then became clear:

1. *Weak elements.* Overselling MRP as a "system," not just the priority planning element, diverted attention from the implementation of master scheduling to drive it and capacity requirements planning to plan the resources needed to support it.

2. *Missing elements.* Almost universally, capacity control was a missing link. Believing that capacity planning was weak, companies failed to utilize input/output control for the major benefits this made possible, even with crude capacity planning.

3. *Oversophistication.* MRP program designers, obsessed with the potential power of computers and software, attempted to build into MRP capabilities to cope with every eventuality in manufacturing and to include every known technique, however little use these would be. Part-Period Balancing with Look-ahead/Look-back lot sizing (see Chapter 6) is a classic example.

4. *Invalid data.* MRP to many users has meant "More Ridiculous Priorities" because of data errors. This subject is discussed in depth in Chapter 8 and mentioned in several others.

5. *Lack of integration.* Data must flow from files in design engineering, process engineering, order entry, purchasing, plant floor, and many other

activities into planning and control files and back out. Transposing these data manually between files delayed the flow, disrupted the timing, and destroyed accuracy. Planning, however well done, was late, unresponsive, and invalid. People could not be held accountable for executing such plans.

It was long known that the steady flow of work on assembly lines was far better for many reasons than batch production in functional work centers with similar processing operations. In the early 1970s, Group Technology (GT) was introduced to improve batch production. This technique grouped together machines and equipment used to make families of parts having similar physical characteristics (shape, weight, material, dimensions) or processing operations (machining, welding, automatic insertion). Advantages included reduced processing cycle times, lower work-in-process, less material handling, and tighter supervision.

There were two significant advantages, lower machine utilization rates and high costs of rearranging plant layouts for changes in processing methods. These tangible costs carried more weight in management thinking than the intangible benefits of faster processing, greater flexibility, and tighter control; group technology was not widely applied in America or Europe.

In the early 1980s, wide interest developed in both America and Europe in Japanese manufacturing practices because of their great advances in quality and productivity. Japan's increasing share of United States markets for autos, motorcycles, electronic equipment, cameras, machine tools, and many other products generated great concern among U.S. competitors. Many study missions to Japan, technical societies, business schools, and politicians attempted to explain how they had made such great gains.

Among many factors, attention centered on Just-In-Time and Kanban "pull systems" and Quality Circles. One effect was to revive interest in Group Technology, newly named Manufacturing Cells, in which one or more machines and operators produce a family of similar parts or products very quickly and flexibly in small lots. Cells, renamed Flexible Manufacturing Systems (FMS), became highly automated as electronic controls were applied to storage systems, material handling equipment, and production machines. Details of these are beyond the scope of this book.

The evolution of these systems made obvious the need for integration of procurement, preproduction, production, and postproduction activities. Design and process engineering, tooling, purchasing, production, quality assurance, and, of course, planning and control people had to become tightly-knit teams to achieve the benefits enjoyed by the best Japanese firms. Even traditional cost-accounting practices had to be changed. All of these are now evolving, too fast for some and too slow for others. Material requirements planning, however, is now a mature technique; no future significant innovations can be expected.

Planning, Control, and Execution

These three terms are very common and are used frequently by many people at all organization levels in manufacturing firms. Their meanings are often different among these people, however. Here are the accepted definitions:

Planning—assigning numbers to future events

Execution—converting plans to reality

Control—tracking execution, comparing to plans, measuring deviations, sorting the significant from the trivial, and instituting corrective actions

Planning and execution have different purposes. They are:

Planning—determining resources needed to carry out plans

Execution—applying available resources to serve customers best

These definitions and statements of purpose make eminently clear why planning and execution require different techniques and systems and that control activities provide interfaces between the two. The planning and control (often misnamed "information") system must support both purposes, recognizing their differences. Two principles apply to planning and control systems:

1. There is one system framework common to all types of manufacturing.

As startling as this is and as unlikely as it seems, it derives inevitably from the common logic, the similarity of needs of the major parties, the nature of the resources employed, and the identical types of data involved. As will become evident when the elements of the system are defined, the importance of each element along with the techniques employed can differ among various types of businesses. The second principle is:

2. There is no single best way to control a manufacturing business.

The second principle may seem at first glance to contradict the first, but it is simply saying that the same vehicle can be used in myriad ways to achieve the different objectives of a number of users with varied interests and to react to changes.

The tools of planning and control, like the weapons of war, include many techniques usable in many ways. Tactics—selecting and applying the right tools—is important for success, but a sound strategy for directing their use is as vital in manufacturing competition as in military battles. A third principle defines the proper basic strategy for manufacturing operations:

3. Do not commit any flexible resource to a specific use until the latest possible moment.

All resources used in manufacturing have flexibility, some more than others. People can handle a wide variety of tasks and acquire new skills. Materials can be made into many different items. Machinery can produce myriad parts and products. Money can buy an almost unlimited variety of goods and services. Fuels and energy have many uses. But once a resource has been consumed in the production or procurement of some specific material, component, product, or facility, its flexibility has been lost.

Plans cover future occurrences, subject to "the slings and arrows of outrageous fortune." They undoubtedly will change, and the amount and effects of changes will be more severe as planning horizons lengthen. The later specific allocations of resources are made, the fewer the changes necessary, the less severe their effects, and the less waste of resources. The strategy just stated derives from the First Law and explains the sources of benefits of faster flows of materials and information.

Each organizational group within a company has its own functions and objectives, but its plans and actions must be communicated to and coordinated with those of other groups through integrated systems. Mislabeled "information systems," these really process only data. *Information is the fraction of such data with specific use by or meaning to people doing their work.* Data processing systems must be designed to produce needed information for users in proper formats.

Planning and execution data are very different. For example, plans show that an order for 50 pieces of a component is to be released week 26, six weeks hence, and is due to be completed week 30. The actual dates it is released and due will probably be earlier or later, and even the quantity can be changed. Planning data, called *soft data,* look precise but often are changed.

If a decision has been made to start an order for one of this component's parents in week 23, this is a firm need date for the component. If work has started on 24 pieces, this too is firm. Execution data need to be precise; they are *hard data.*

At the heart of all manufacturing planning is a core system, shown schematically in Fig. 1-2, linking the several functions involved. This framework embodies the same fundamental logic and applies the methodology of techniques commonly used by practically every manufacturing company. The universal system principle stated earlier derives from these facts.

In such computer-based systems, data are grouped in "files" with names like unshipped customer orders, bills of material, approved suppliers, released purchase orders, unfinished work orders, and product costs. These files have similar names, sources of data, and uses in all manufacturing companies. The data, of course, are specific to each individual firm, but the file structures and processing logic can be identical. The basic requirement for integrated systems and networks of related systems is that the structures of files permit computer updating, accessing, and downloading of data without manual intervention.

In Fig. 1-2, the Core System links Strategic, Business, and Production Planning, likely to be quite different in individual companies, although the

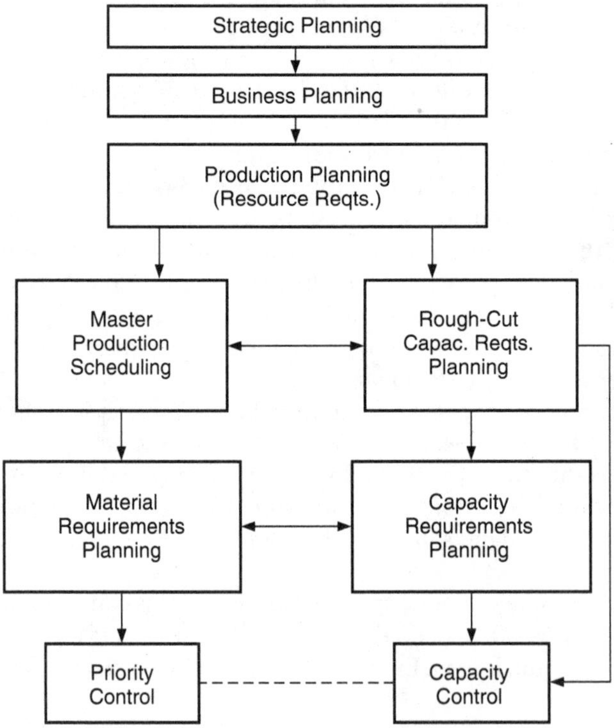

Figure 1-2. The core system.

types of data handled are common and include forecasts, product designations, and facilities identity.

Strategic plans cover long horizons and include a statement of the charter of the business, a brief summary of the type of business it is and will become. They also include strategies for the marketplaces entered, types of products, and manufacturing practices. *Business plans* have a shorter horizon and focus on product families to meet strategic goals.

Production plans cover manufacturing facilities needed to support both strategic and business plans. They specify quantities of families of products and amounts of resources in plant and support facilities over the same horizon as business plans. Resources include capital, people and skills, plant and equipment, make-or-buy decisions, and other factors defining the facilities needed. They are more detailed than strategic and business plans and may cover a shorter horizon.

The structure of the core planning system integrates five basic elements utilizing techniques common to all manufacturing:

1. *Master production scheduling (MPS)*—developing the data showing which products should be produced, how many, and when. These data are constrained by the Strategic, Business, and Production plans and drive all detailed operating plans.
2. *Material requirements planning (MRP)*—determining quantities of materials needed to support the MPS and when they should be scheduled. The terms *material planning* and *priority planning* generally are used interchangeably in MRP.
3. *Capacity requirements planning (CRP),* rough-cut and detailed—evaluating the resources (people, machines, capital, etc.) needed to support the MPS. For longer-range planning, rough-cut techniques are simpler, faster, and better. A detailed discussion of these techniques may be found in Chapter 9.
4. *Production activity control (PAC)*—after the planning phase, actual performance in the material/priority execution phase is compared to plans and significant deviations highlighted for corrective actions. This is covered in Chapter 10.
5. *Input and output capacity control (I/O)*—actual capacities are compared to plans and signals sent for actions needed. See Chapter 10 for a detailed discussion of this also.

Manufacturing planning and control involve three functions:

1. *Developing a sound game plan.* This requires two sets of numbers: first, a Production Plan (generalized long-term) and, second, Master Production Schedules (specific short-term).

2. *Adequate capacity* in resources needed to support the game plan, answering, "Will enough in total be produced?"
3. *Correct priorities,* answering the question, "Are the right things being worked on in the right sequence?"

The third question is answered by the core system's priority elements and the second by capacity elements. For any manufacturing business to be in good control, both answers must be "Yes!"

Priority and capacity elements are very different and distinct, but they are inseparable. A fourth principle is

4. Priority information and material schedules cannot be valid, however desirable, if resources are or will be inadequate; such plans are impossible to execute.

Planning can be done by computer-based systems with minimum human involvement, but successful execution depends on people taking the proper actions promptly. Effective planning systems are necessary but not sufficient; they make tight control *possible,* but people make it *happen.*

2
Inventory Management

If you can't control what is put in and taken out, you can't control an inventory. If you can, you don't need the inventory.

Inventory Classes and Functions

Manufacturing companies have two classes of inventory:

1. Manufacturing inventory, consisting of
   a. Raw materials
   b. Semifinished component parts
   c. Finished component parts
   d. Subassemblies
   e. Component parts in process
   f. Subassemblies in process
2. Distribution inventory, made up of
   a. Completed products in warehouses
   b. Completed products in transit

Production inventories have five separate and distinct functions. The functions and class names are

1. *Decouple manufacturing operations*—different facilities (people and equipment) process materials at different rates in a time period and cannot be linked rigidly. Batch order quantities, called *lot-size inventory,* separate them.

2. *Cushion against upsets*—demand changes and interruptions in supply have costly and harmful effects. Buffers, called *fluctuation inventory,* minimize these effects.
3. *Level production*—cyclical and seasonal changes in demand are expensive (often impossible) to handle, and early production for anticipated changes is necessary. The stocks built for these reasons are called *anticipation inventory* or *stabilization stock.*
4. *Fill distribution pipelines*—materials in transit are called *transportation inventory.*
5. *Hedge against expected events*—suppliers' price increases, labor strikes in suppliers' facilities or in transportation, new government regulations, and similar events may make *hedge inventory* a good investment.

The functions of a system for managing any inventories, even those of a grocery, museum, or blood bank, as well as a manufacturing business, are

1. Planning
   a. Setting policies
   b. Forecasting demand
   c. Selecting and using techniques
2. Acquisition
   a. Positive order action (place or increase)
   b. Negative order action (decrease or cancel)
   c. Receiving
3. Stock keeping
   a. Physical control
   b. Accounting (record keeping)
4. Disposition
   a. Purging (scrap and write-off of obsolete items)
   b. Disbursement (delivery to source of demand)

Any system of inventory management can be described this way functionally. Production inventory management, however, has its own distinct characteristics, compared with nonmanufacturing, showing differences in the content of key functions in every one of the four principal activities just mentioned. For example:

1. *Planning.* Normally, policy for production inventories has an objective of always having the least inventory consistent with good customer service, meeting production requirements, and holding manufacturing costs to a minimum. Forecasting within the production inventory system

plays a minor, secondary role compared to estimating demand from outside sources. Because most demands for components and raw materials are dependent on production decisions, techniques used are different.

2. *Acquisition.* The order action function is expanded, and exhibits several characteristics unique to production. Materials, from the inventory system's point of view, are being acquired, dispersed, and reacquired as they progress through multiple stages of conversion from raw material to end products.

An order for a manufactured item, once initiated, may incur the penalty of scrap and rework. Normally its quantity cannot be increased; decreases are possible but raise overhead costs. The ordering function includes order suspension (rescheduling to an indefinite future date). Finally, the quantity and timing of an order may be affected by available capacity of a plant and its suppliers.

3. *Stock keeping.* Inventory accounting functions may be integrated with or merged into inventory planning functions.

4. *Disposition.* Delivery of a production inventory item is always to an in-house demand source, including processing facilities and stockrooms, to meet a production schedule. When an inventory item is received from a supplier or has been completed, it is earmarked for use in the next processing stage or it enters a distribution inventory.

In a manufacturing environment, inventory management is inseparable from production planning. The function of inventory planning is to translate the overall plan of production (the master production schedule) into detailed component requirements and orders. This determines, item by item, what is to be procured, how much, and when, as well as what is to be manufactured, how much, and when. Its outputs "drive" both purchasing and manufacturing activities, providing requisitions or orders authorizing them. The inventory program recommends order priorities and makes capacity requirements planning possible, assuming that adequate capacity will be available when required. It does considerably more than merge inventory; it is the heart of logistics planning.

The purpose of distribution inventories like those found in supermarkets' or wholesale distributors' or in manufacturers' finished-goods or service-parts warehouses is to supply quickly the items customers demand. Such demands usually are difficult to forecast; they tend to be erratic, typically made up of small demands originating from many separate sources. The inventory investment level is governed by marketing considerations such as competitors' performance and shipping costs and delays.

The purpose of production inventory is quite different; it is to satisfy production requirements. Component and raw material availability are geared to a production plan, meaning that demand is calculable. Demand for an item in any period typically consists of a limited number of individual demands for multiple quantities. Except for components for field replacement, production is the sole source of demand and it is always finite. The inventory investment level is determined by manufacturing considerations like setups and processing rates. Work-in-process, an inventory entity unique to manufacturing, may constitute a significant part of the investment; the level of this inventory is primarily a function of planned manufacturing lead times. Often the responsibility for finished products passes to a marketing, distribution, or service organization once production is complete.

The difference between distribution and production inventories is fundamental. Consequently, the respective inventory management policies, systems, and techniques used are (or should be) fundamentally different. In determining the desirable level of a distribution inventory, the trade-off is between shipping, warehousing, and carrying costs of the inventory investment and added sales revenue realized through improved availability to customers. Theoretically, 100 percent service requires an infinitely large inventory investment—in practice, there is some large amount that avoids stock-outs.

In determining a production inventory level, there is no such trade-off. The investment is dictated by production requirements, which, unlike customer demand, are known and controllable. Inventory that exceeds the minimum required brings no extra revenue. A 100 percent service level for component items used to make finished products is feasible with a finite investment.

In a distribution inventory environment, demand for each inventory item must be (explicitly or implicitly) forecast. Uncertainty exists at the item level. The principle of stock replenishment (to restore availability) applies, and the two principal questions are when to reorder and in what quantity. The first cannot be answered with certainty. The second is answered through the computation of some economic order quantity or economic warehouse replenishment frequency.

In a production inventory environment, individual item demand can be calculated and need not be forecast. Uncertainty exists only at the master production schedule level. There is no need to replenish depleted inventory immediately, only to get whatever is required at the right time to cover future production needs. In-stock inventory indicates premature availability. Ideally, all production inventory would be in process, with every item immediately used in the next manufacturing operation upon receipt from suppliers or completion of processing. The best-managed manufacturing inventories now approach this ideal.

Reappraisal of Concepts

With the advent of material requirements planning, many of the existing concepts and techniques of inventory management needed to be reappraised. Questions were raised about the validity, relevance, and applicability to production inventories of these commonly accepted ideas:

1. The concept of stock replenishment
2. All techniques involving reorder points
3. Square-root economic order quantities
4. Classification of inventory by function
5. Aggregate inventory management concepts
6. Controls based on ABC inventory classification

The term *stock replenishment* means restoration to a state of prior fullness, but production inventories should, if possible, be handled in exactly the opposite way. The concept of stock replenishment, therefore, is in conflict with one of the three basic management objectives:

1. High customer service for high sales revenue
2. Low inventory for high return on investment
3. Low costs for high profits

Stock replenishment attempts to have inventory items in stock at all times, in spite of poor predictions of both amount and timing of need. In manufacturing, of course, the ideal is to have inventory items available just at the time of need, not sooner or later. Near-term needs (both quantity and timing) for individual production inventory items can be determined through the use of modern, computer-assisted MRP programs. The excess inventory associated with stock replenishment techniques is undesirable and wasteful.

Techniques built on *order points* (OP) implement the stock replenishment concept. They include statistical order points, min/max, order-up-to targets, two-bin, periodic review, and the maintenance of *N* months' supply. Explicitly or implicitly, all use forecasts of demand during replenishment lead time and all attempt to provide safety stock to compensate for fluctuations in demand. Programs based on reorder point techniques suffer from false assumptions about both demand and supply environments, misinterpret observed demand behavior, cannot determine the specific timing of future demands, and lack the ability to be replanned. These shortcomings cause unnecessarily high inventory levels of some items, shortages of others, and inventory imbalances of unmatched sets of parts. These techniques are discussed in more detail later in this chapter.

Square-root economic order quantities are not economical for typical manufacturing demands, because the calculation does not relate the lot size to either the timing or the quantity of actual, discrete demands (requirements) arising during specific time periods. EOQ calculations are based on ordering and inventory carrying costs, neither of which can be determined rigorously. They assume long periods of usage and several other fallacies. These techniques are covered in Chapter 6.

ABC inventory classifications are an adaptation of Pareto's Law of the vital few and trivial many. In any inventory (and in many other groups in manufacturing) the vital few comprise roughly 20 percent of the items but account for 80 percent of the value; in a typical ABC classification, these are designated as A-items. Of the remaining 80 percent of the items, typically 30 percent, accounting for 15 percent of the value, become B's and the remaining 50 percent are C's, representing only 5 percent of value. The idea behind ABC classes is to apply the bulk of limited planning and control resources to the A-items "where the value is," at the expense of the other classes that have demonstrably much less influence on the total. The ABC concept is implemented by controlling A-items "more tightly" than B-items, and these better than C-items.

The Logic of Manufacturing

The fundamental logic of manufacturing—What will we be making? How many of each component are needed? How many do we already have? When do we need the rest and how will we get them?—has been used since cave dwellers made slings, bows, arrows, and spears. In pre-MRP industry, the first question was answered using forecasts of future demand, unless a large backlog of customers' orders was available (the rare exception). The next three questions required great amounts of detailed information on products, inventories, and processes so often lacking integrity that crude estimates and approximations were substituted.

Manufacturers of large, complex equipment (for example, ships, trains, planes, central station boilers and generators) had long future horizons covered with firm orders. Planning was manual, slow, and crude. Large clerical groups calculated gross requirements for major components of their products and time-phased (albeit very roughly) these and their procurement. Revising such plans was even more tedious and was rarely done. The capability of massive data storage and manipulation required for sound inventory planning simply did not exist at that time.

Because of this constraint, stock replenishment (OP/EOQ) methods predominated prior to the 1970s. Inventory control was attempted utilizing paper records and electromechanical desk calculators to apply essentially simple mathematical formulas for order-quantity and safety-stock calculations.

Part of the last question above, "How many should we buy or make?", was answered in the decades preceding MRP with EOQ techniques. As mentioned in Chapter 1, Ford Harris published the first theoretical formula for EOQ in 1915. In early practice, the question of the "correct" order quantity deserved and received only secondary attention. This question does not arise at all when the demand for an inventory item is either highly continuous (typical for high-volume production operations) or very intermittent. It is obviously more important to *have the quantity needed at the time it is needed* than to order an economic quantity. Evidence of this was frequent splitting of lots in process, double and triple setups caused by "hot-order" expediting, and partial vendor shipments, all of which were normal occurrences.

The first half of the last question, "When are raw materials and components needed?", received the crudest of answers prior to computer-based MRP programs. Statistical calculations of safety stocks, proposed by R. H. Wilson in 1934, gave the appearance of precision without the reality of accuracy. Calculated EOQ and safety stocks, and subsequent refinements and elaborations, did improve production inventory control over the guesstimates and estimates preceding them, but left much to be desired.

Dependent and Independent Demand

Orthodox inventory analysis and classification techniques are designed to determine the most desirable treatment of a given inventory item or group of items. They employ various attributes of the items such as cost, lead time, and past usage, but none of them takes into account the most important one, the nature of demand. Yet it is the nature (or source) of demand which provides the real key to inventory control technique applicability and selection. The fundamental principle in the selection of order point or material requirements planning is *whether demand for the item is dependent or independent of that for other items.*

Independent demand must be forecast, but dependent demand can be calculated since it is directly related to, or derives from, the demand for another inventory item or product. Dependency may be "vertical," when a component is needed in order to build a parent subassembly or product, or "horizontal," for an attachment to a product or an owner's manual shipped with the product. In most manufacturing businesses, the bulk of the total inventory is in raw materials, component parts, and subassemblies, all largely subject to dependent demand. Dependent demand need not and should not be forecast; it can be precisely determined from the demand for those items that are its sole cause.

OP/EOQ systems are oblivious to the dependence of one inventory item on another. Order point views every inventory item as though it had a life of its own; thus, forecasting and OP are inseparable. But all forecasting (intrinsic or extrinsic, judgmental or statistical) attempts to use past experience to determine the shape of the future and succeeds only when the future continues to be like the past.

In a manufacturing environment, however, the objective is to make the future better than the past; future demand for a given part will almost invariably be different from past demand. Marketing and sales people and competitors are working diligently to make it different. Forecasting, therefore, should be the method of last resort, *never used when it is possible to relate one item's demand to that for another item.*

In manufacturing, an inventory item may be subject to dependent demand exclusively, or to both dependent and independent demand. Such mixed demand arises in cases of parts used in current production as well as sold for spare parts. The independent portion of the total demand (spare parts sales) has to be forecasted and added to the dependent demand (to make a parent item). Service parts no longer used in current production, and obsolete components still sold as products, are subject to independent demands exclusively; these are properly forecasted.

Production operations convert raw materials into component parts, which go into subassemblies and then into finished products. In many plants, semifinished parts and several levels of subassemblies are made. The typical relationships among manufactured inventory items are depicted in Fig. 2-1. Purchased steel is forged into a blank and this is machined into a gear, which then joins other components assembled into the gear box, which goes into a transmission. This in turn will be assembled into some vehicle, sold, and shipped to a customer. Each item carries a unique identity (part number) and must be planned and controlled. Demand for the vehicle may (probably will) have to be forecasted, but none of the component items requires an individual forecast; their demand is created internally by planning the next conversion stage and can be calculated.

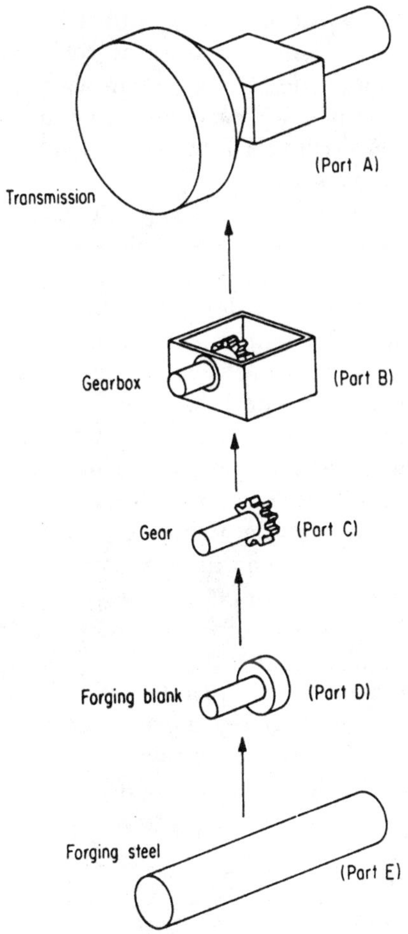

Figure 2-1. Material conversion stages.

Order Point versus MRP

There are two alternatives and two associated sets of techniques available for managing production inventories:

1. Order point (OP)
2. Material requirements planning (MRP)

Order point, also called *satistical inventory control* and *stock replenishment,* is defined as a set of records, procedures, and decision rules intended to ensure continuous availability, even with uncertain demand, of all items comprising an inventory. Using OP, the depletion in the supply of each inventory item is monitored and a replenishment order is released whenever an issue drops the supply to a predetermined quantity—the (re)order point. This quantity is determined for each inventory item separately, ignoring its use in assemblies, using a forecast of demand during replenishment lead time plus safety stock.

Material requirements planning consists of a set of records, logically related techniques and procedures, and decision rules to translate a master production schedule (MPS) for finished products, major subassemblies, or end items into time-phased net requirements and orders for each component needed to implement this schedule. MRP replans net requirements and coverage for changes in MPS, design, processing, or inventory status.

Requirements for items stated in the MPS are derived from forecasts, customer orders, field warehouse requirements, and interplant orders. Developing and applying this vital set of data is covered in Chapter 8. Requirements and timing of orders for all components are determined from MPS by MRP. The term *component* in material planning covers all inventory items other than finished products, often called *end items.*

Order point is part-based; material requirements planning is product-oriented. OP utilizes data on the historical behavior of an inventory item's demand, in isolation from all other items. MRP ignores history, looking toward the future as defined by the MPS and works with data specifying the relationships of components (the bill of material) of a product. Other differences between OP and MRP are explained later in this chapter.

With two alternatives available for production inventory management, three questions arise:

1. Are they mutually exclusive or compatible?
2. Which is preferable under what circumstances?
3. What are the principal criteria to be used in making a selection of which to apply?

Before the answers to these three questions can be understood fully, the concept of dependent and independent demand just discussed must be applied.

Order Point Characteristics

Order point theory makes five basic assumptions:

1. Independent demand can be forecast with reasonable accuracy.

2. Such forecasts will account for all demands.
3. Safety stocks will protect against forecast errors and unexpected events.
4. Demand will be fairly uniform in the short-term future, and a small fraction of reorder quantities.
5. It is desirable to replenish inventories when they are depleted below the order point quantity.

Forecasts of demand for components in manufacturing are most often derived from each item's past usage (intrinsic forecasts), rarely from finished product or other external demand (extrinsic forecasts). Very few people try to forecast separately demand from each product for an item common to several products. Demand forecasting determines only the average amount of demand expected in future time periods, not the timing of specific demands.

Everyone who has used forecasts knows that they will be wrong; the only question is, "How wrong?" Order point attempts to protect users against forecast errors and other unexpected happenings by adding a cushion called *safety stock*. It uses some economic order quantity computation to specify the size of the replenishment order.

When random-access computers and applicable software became available, more sophisticated applications of order point could track actual demands and compare them to forecasts (also updating these periodically) to indicate the probability of actual demands exceeding the forecast. Statistical techniques could then be applied to calculate an amount of inventory (usually called *safety stock*) which would assure achieving a desired level of "service," meaning some minimum number of stock-outs.

How well order point inventory planning works depends on how closely the assumptions relate to the actual situations in the inventory. Introduced with manual calculations in the 1940s, order point systems were an improvement over the earlier crude "guesstimates" approaches. The enhancements made possible by computers further improved their performance, but the fallacious basic assumptions defeated their users in getting tight control over manufactured inventories.

The OP assumption of fairly uniform usage in small increments is invalid in a manufacturing environment. Requirements for components of products are anything but uniform; depletion occurs in discrete "lumps" caused by parent order lot sizes. The example in Fig. 2-2 shows this clearly; in it, order point is being used for all items. These could be a box wrench, the rough forging it is made from, and the forging steel. Wrenches and more complex products are not made in quantities of one piece, of course, but in reasonably-sized lots.

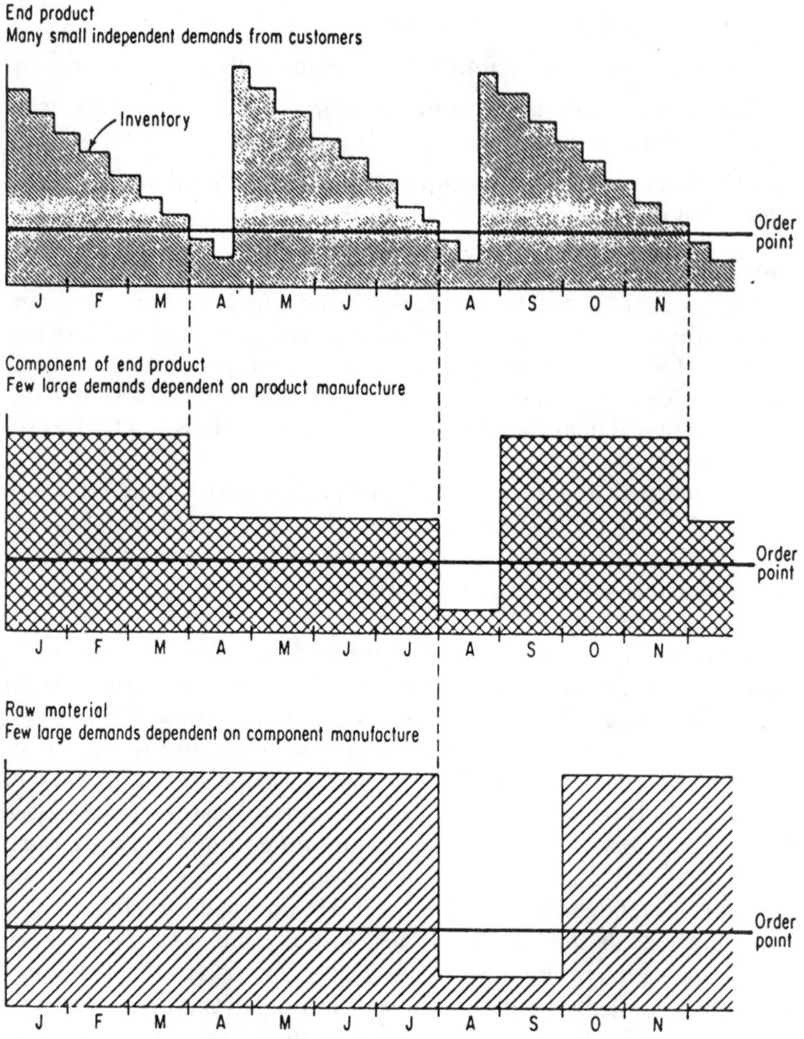

Figure 2-2. Order point and dependent demand.

When an order is placed on the factory to produce a quantity of such an end item, it is necessary to withdraw from inventory a corresponding quantity of the component; this will deplete its inventory and at some time drive it below the order point. When it does (as at the end of July, in the example), the technique will act immediately to reorder it, necessitating a large withdrawal of raw material to produce its order quantity. If the raw material order point is then "tripped," this material will be reordered immediately also.

In this example, demands for the component and raw material show marked discontinuity, causing several serious problems:

1. Average inventory levels are considerably higher than one-half the replenishment lot size plus safety stock, which OP/EOQ theory commonly assumes.
2. The order point system reorders prematurely, far in advance of actual need, and excess inventory will be carried for significant periods of time.
3. The schedule dates on the replenishment orders are wrong, and credibility of the system will be low.
4. Scarce capacity and materials will be applied to the wrong items.

The example in Fig. 2-2 shows graphically the effects on the performance of OP systems of the fallacies on which they are based: demand is not continuous and not in small increments, more inventory is not needed immediately, orders are released too soon, and even large safety stocks give dubious protection.

The wrench, forging, and steel in the preceding example can be used to demonstrate other reasons for poor performance of OP systems. Figure 2-3 shows four types of wrenches, two (A and B) made from one forging (X) and the other two (C and D) made from a different forging (Y). Both forgings are made from steel (Z). In each of the simplified records, the letter D stands for demand and P for production.

To make one wrench requires one forging made from one unit of steel, but each has its own production lot size. Forgings will be needed every time

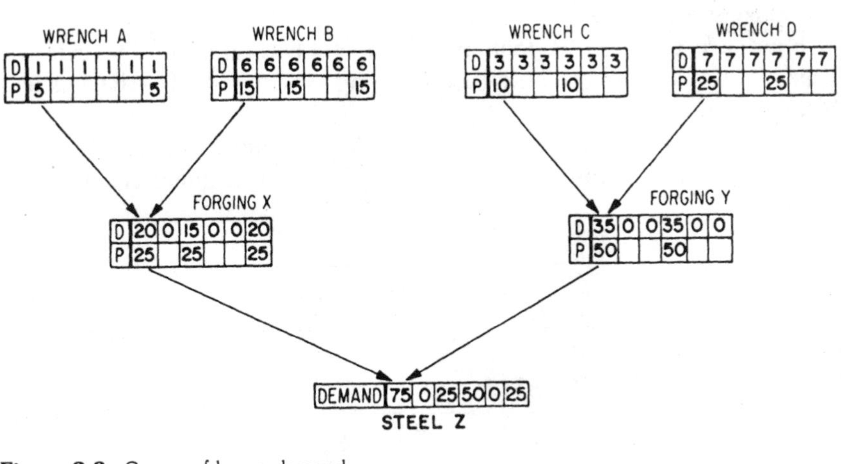

Figure 2-3. Causes of lumpy demand.

a batch of wrenches is produced; the demand will be the wrench order lot size. Every time forgings are made, steel will be required for the forging batch. The lot sizes for each component are shown on the P line, and it is assumed that the quantities will be ready when needed.

Demand for each finished wrench in the example is shown to be uniform. This is hardly realistic, but was chosen to highlight the fact that even continuous and uniform demand for end products can become intermittent, very lumpy demand for components. The size and timing of the lumps, of course, are functions of the period chosen. By increasing the size of the period in our example from a week to a month or more, assuming the wrench requirements are still valid, even the demand for item Z, the steel, can be smoothed; over a 50-week cycle, its demand will be a uniform 850 units. Smoothing over long periods may be helpful in capacity management but will be useless and even harmful for material planning. Manufacturing inventory and production must be planned, scheduled, and controlled from day to day.

In our example, the total demand for wrenches in each period is 17 (=1 + 6 + 3 + 7); the average demand for steel Z is also 17 per period, since one unit of steel is consumed in the production of each wrench. Because of its assumptions, order point always contains a planned safety stock; 17 units, representing one week's supply, seems reasonable with a 4-week lead time. The basic formula for order point is

$$\text{Order point} = \text{Demand during lead time} + \text{safety stock}$$

For our example,

$$\text{OP} = (17 \times 4) + 17 = 68 + 17 = 85$$

There are 180 units of steel on hand now. Lead time for new orders is 4 weeks. Actual demands and the remaining available steel each week will be as shown in Fig. 2-4.

Order point tracks usage *as it occurs*. In Week 4, when an issue of 35 pieces drops available inventory below order point, OP triggers release of a

WEEK	1	2	3	4	5	6	7	8	9	10
DEMAND	65	0	0	35	30	0	35	0	0	35
AVAIL	115	115	115	80	50	50	15	15	15	–20

Figure 2-4. Steel requirements and availability.

replenishment order. *Lead time is added* to set the due date at week 8. OP knows nothing of future needs.

Order Point Plan: Start week 4; due week 8.

MRP *calculates future requirements and available inventory* but *plans no safety stock*. It sees that a new order will be needed in week 10 when available inventory is all gone and sets its release for week 6 by *deducting the lead time*. It is not concerned with past events but looks only to the future.

MRP Plan: Need week 10; start week 6.

Why the difference in order dates? Is it caused by the 17 units of safety stock used in OP but not in MRP? If MRP planned this safety stock, it would see a new order needed in week 7 (to protect the safety stock) and started in week 3, still different from OP. Which is right for this item? Wrong question! The right question is, Which is better for *many* such items?

This example shows only some of the differences between OP and MRP planning; all are summarized in Fig. 2-5.

How does each work when matched sets of parts are needed for assemblies? Order point was used by one company making an assembled product from 8 components. Over a year, the "service level" for these items, measured by number of weeks out of stock divided by 50, looked good; it ranged from 92 to 100 percent, as in Fig. 2-6.

The figure shows also the (relatively few) weeks each item was not available; obviously, finished products could not be produced in these 11 weeks (number 35 appears twice). The assembly department saw only 78 percent service, not the high 90s percentages.

	Order point	MRP
Deals with	parts	products
Looks to the	past	future
Uses	averages	batches
Requires bills of material?	no	yes
Inventory is	maintained	run out
Recommends order dates to	start	complete
Activated for an order	once	periodically
Shows future orders	none	all in horizon
Can be replanned?	No	Yes

Figure 2-5. Comparison of ordering techniques.

Part #	Wks out of stck	Service level, %	Wk #'s out of stock
1	3	94	11, 12, 13
2	1	98	50
3	2	96	30, 35
4	0	100	
5	1	98	42
6	4	92	23, 24, 35, 36
7	0	100	
8	1	98	19

Figure 2-6. Component service levels.

In one week in which all items were available, the inventory balances of each item were:

Item	Inventory	Item	Inventory
1	81	5	226
2	719	6	34
3	1134	7	87
4	193	8	349

One of each item is needed to make an assembly. Limited by item 6, only 34 finished products could be built! The large excesses of other items' inventory over this total were useless. The explanation, "They resulted from using EOQs," was only part of the answer. They were caused by using the wrong ordering technique. Most products have many more than six components.

If the probability of having one item in stock when needed is 90 percent, two items controlled independently but needed simultaneously have a combined probability of only 81 percent (0.9 × 0.9). Ten items will all be available only 34.8 percent of the time. Even with desired service levels for components set at 95 percent, this probability is still less than 60 percent. Table 2-1 shows these probabilities for 90 and 95 percent component service levels for up to 25 items per assembly; many products have 10 or more times this many. Failure to distinguish dependent from independent demand and to select the proper technique can be disastrous.

No wonder OP techniques applied to manufactured components cause lots of expediting! They are rational methods for getting into difficulty when misapplied. They have been described as "steering the ship by watching the wake," with only the hope of adequate safety stocks to protect against hazards ahead.

Table 2-1. Probabilities of Simultaneous Availability

Number of component items	Service level 90%	Service level 95%
1	0.900	0.950
2	0.810	0.902
3	0.729	0.857
4	0.656	0.814
5	0.590	0.774
6	0.531	0.735
7	0.478	0.698
8	0.430	0.663
9	0.387	0.630
10	0.348	0.599
11	0.313	0.569
12	0.282	0.540
13	0.254	0.513
14	0.228	0.488
15	0.206	0.463
20	0.121	0.358
25	0.071	0.260

A variation on OP is the Periodic Review technique, also called *fixed-cycle ordering* or *sales replenishment,* in which inventory records are reviewed periodically and replacement orders issued. This technique is applied when

1. There are many small issues from inventory and it is impractical to post every one—typical of retail businesses.
2. Ordering costs are low, as they are when picking items in a central warehouse for branch warehouse deliveries.
3. It is desirable to order several items simultaneously to run families in economical production.

Figure 2-7 is typical, showing equal review periods with a target level above the sawtooth pattern of inventory. The target (order-up-to amount) is set equal to the sum of projected demand during lead time (DLT), safety stock (SS), and projected demand during the review period (DRP). The first two are identical to factors in OP; the last covers the additional time of the review.

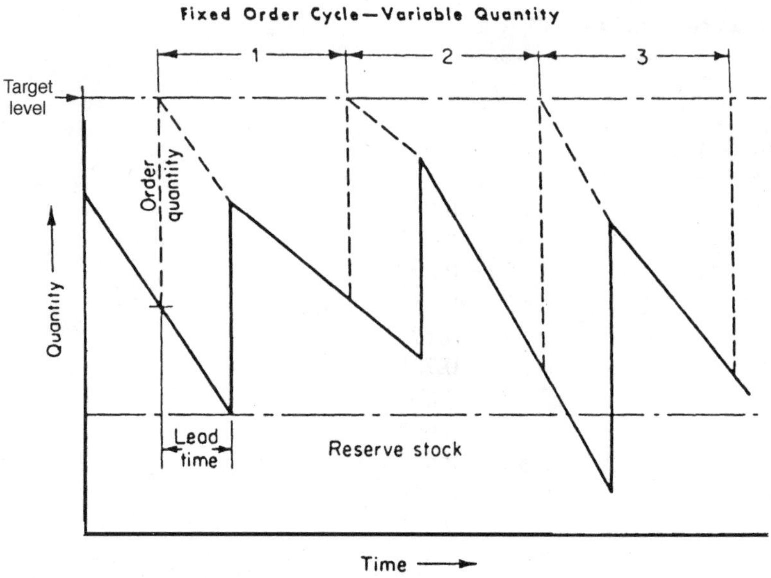

Figure 2-7. Periodic review system.

An example of the target calculation is

Forecast demand = 20 units Safety stock = 30 units

Lead time = 1 period Review cycle = 2 periods

$$\text{Target} = \text{DLT} + \text{SS} + \text{DRP} = 20 + 30 + 40 = 90 \text{ units}$$

Using this technique,

1. Total lead time is actually equal to replenishment lead time (at the source) plus the review period.
2. Longer review periods require higher safety stocks.
3. It is desirable to order several items simultaneously to run families in economical production.

This technique is excellent for replenishing multiple branch warehouses from a central source; review periods can be scheduled to spread the load on shipping people, get regular deliveries at warehouses, and keep inventories in balance. It also helps reduce paperwork of ordering, invoicing, and paying for deliveries.

The technique called *time-phased order point (TPOP),* described in Chapter 4, improves the performance of OP in managing inventories of items with independent demand. TPOP applies MRP logic to the independent portion of demand for such items. The essential difference from MRP is forecasting an item's demand instead of calculating it from the plan for its parent(s). TPOP, like MRP, provides forward visibility of replenishment needs and orders and, unlike OP, does not depend on actual issues to trigger new orders.

Simple ordering methods, used commonly for low-value C-items, include

1. *Min/max,* using estimates of minimum stocks to cover reorder periods and estimated order quantities. When stocks reach "Min" levels, orders are placed for the difference between actual stock and "Max" (equal to Min + OQ).
2. *Maintain N periods,* by ordering up to *N*-periods' worth when supply reaches an estimated replenishment amount.
3. *Two-bin method* places a quantity (equal to the OP quantity) in a separate storage bin or container, not to be touched until the working stock is used up. When accessed, a replenishment order is placed.
4. *Visual review method* periodically checks stocks of items against crude levels for each item (½ bin, or 8 inches deep). When stocks reach these "reorder points," replenishment orders are placed.

These techniques involve a minimum of record keeping; the last two depend on physical actions rather than system data. All require tight discipline in following the prescribed methods. They must be checked also for significant changes in usage rates or replenishment lead times. The objective is to avoid shortages by having plenty that doesn't cost much.

Mixed Independent and Dependent Demand

While MRP is intended primarily to handle components having dependent demand, it can easily accommodate mixed dependent and independent demand items. Service parts used as components in current production are an example. When the independent demand is forecast, the TPOP technique can be utilized to project its time-phased inventory plan for future replenishment of service part inventory. The MRP program then simply adds these requirements, in the proper time periods, to the component's dependent demand, calculated from end-item requirements stated in the MPS. In addition to linking both types of demand in one system, TPOP also provides replanning capability for independent demand items such as finished products in factory and field warehouses.

Material Requirements Planning

Material requirements planning is the preferred technique to use when one item is a component of another, thus subject to dependent demand. MRP does not rely on a forecast of item demand and does recognize sets of items, thus avoiding two major deficiencies of OP. This technique is expressly designed for dealing with dependent, discontinuous, nonuniform demand, characteristic of most manufacturing environments. It is forward-looking, not history-driven like OP. It develops a valid inventory plan that can be replanned to keep it up-to-date. MPR is not without its own problems, of course. These will be understood better after the mechanics of MRP and its related techniques, covered in Chapter 4, are known.

When MRP was introduced, revolutionary changes occurred and new premises were established. Orthodox practices and techniques were challenged. The existing inventory-control literature, indeed an entire school of thought, had to be reexamined. It became evident that the basic tenet of old inventory-control theory—that low inventory investment and high customer service were incompatible—was mistaken; successful users of MRP programs reduced inventories and improved delivery service at the same time.

Notwithstanding the difficulties experienced by many companies attempting to implement MRP-based planning, MRP has been the most successful innovation in manufacturing inventory management. It has demonstrated conclusively its operational superiority over alternatives in many companies in many industries in many countries. In addition, it has provided students of inventory management with new insights into manufacturing planning and control. MRP shows the true interrelationships among activities affecting inventories and reveals the fallacies of previous concepts, pinpointing the causes of inadequacies of older methods.

The Paradox of Inventory Management

A remarkable truth about inventory management is

> **If you cannot match inputs to and outputs from an inventory, you will never control it.**
>
> **The better you can match inputs and outputs, the less need there is for any inventory.**

Inputs to inventory are receipts of purchased materials from suppliers and completed production of parts, subassemblies, and assemblies. Outputs are deliveries to customers, external and internal, issues to manufacturing processes, and scrap.

People concerned with inventory and its management include top-level managers, sales and marketing, purchasing, materials planning and control, production activities, and cost accountants. Until very recently, most of these people were involved with outputs or inputs separately or thought of them individually, not recognizing their interactions. They believed they were victims of the environment.

Many believed that Murphy's Law, *What can go wrong, will go wrong at the least opportune time,* was written about factories. Problems abounded. Customers changed their minds about what they wanted, how many, and when they wanted them. Designs were faulty and changed frequently. Suppliers were late and often delivered defective materials. Machinery and tooling broke down. People were absent and made defective things when they worked. The consensus was that these problems were unsolvable.

This has now been proven false. Most can be solved and eliminated; those that cannot be eliminated can be minimized. There is now little excuse for the "surprises" blamed on manufacturing problems in the past.

A major objective of this book is to show the important role MRP can play in helping to achieve a better balance between inventory inputs and outputs. The paradox stated above has made it clear how we should answer the question, "How much inventory is enough?" The answer is, "How much is needed because inputs and outputs don't match?" Even more important questions are, "Who is responsible for the mismatch?" and "How long will it take to get them to match?"

Inventory—Asset or Liability?

Management has always challenged amounts of manufacturing inventories—raw materials, work-in-process, and stocks of component parts and subassemblies. Although they called them assets (and they comprise a major portion of the asset total on most companies' balance sheets), executives and top-level managers have viewed and treated them as liabilities. Improving inventory turnover (Total Sales at cost, divided by Total Inventory at cost) has been the perennial goal of management. Only finished product inventories have been accepted, albeit grudgingly, as necessary to serving customers.

This abhorrence of inventory now is seen as being eminently correct—for different reasons! In itself, some inventory is needed in manufacturing;

some is even beneficial, earning an adequate return on the investment. Such inventory is an asset, but in most firms it is a very small fraction of the total.

The single most important performance measure of the overall health of a manufacturing business is inventory turnover ratios. In the United States before 1980, these ranged from below 1 to as high as 6. Management thought they were doing well to increase the figure by 50 percent in one year. Many who did so simply fell back to previous ratios the next year, indicating successful crisis actions but no permanent improvement in performance. Many found that lowering inventories harmed customer service and caused higher costs, indicating lack of understanding of how manufacturing should work and failure to use sound planning and control.

Harold Geneen, the CEO and president who built the ITT corporation into a massive conglomerate and led it to 14 years of increasing profits, is reported to have said to one of his division presidents, "You told me two years ago that inventory went up because of the boom in sales and your struggle to catch up quickly. You now tell me inventory went up because sales fell off and you couldn't shut off materials. When does inventory ever go down?" He knew the answer. It goes down when top management demands, "Take it down!"

Issuing the edict is necessary, but not enough. The people responsible for acting on it must have the knowledge and tools to react properly to avoid damaging customer service and profits by instigating crisis actions. This must begin with sound planning and end with tight control. MRP is the core of modern integrated planning and control systems. It is the key to effective inventory management.

3 Prerequisites of MRP

If we had ham, we'd have ham and eggs, if we had eggs.

Perspective

The alternative choices of order point and material requirements planning for inventory planning have been presented in the preceding chapter. OP weaknesses in a manufacturing environment are highlighted. MRP is a technique that recognizes the realities of demand existing in this environment. MRP, however, makes some assumptions and requires information about characteristics of the product and of the process used in its manufacture. These assumptions and prerequisites of MRP and the principles on which it is based are the subjects of this chapter.

Over the last three decades, material requirements planning has been evolving steadily, moving toward higher capabilities with every enhancement in techniques and data processing power. It has been painstakingly developed into its present stage of sophistication by practicing inventory managers and planners.

Order point never made sense to thinking practitioners for thousand-dollar components, for instance. Why stock them, and reorder them in economic lot quantities, when they could determine by simple calculations how many of each would be required when, using the master production schedule? They treated expensive items in the same way MRP treats all items under its control. Specific demand for items is calculated rather than forecast; any existing surplus is taken into account to determine net demand and the date more are needed. Lead times are deducted to determine order release dates, items are ordered discretely, and detailed surplus records are kept.

The evolution of MRP was equivalent to applying the strict and careful treatment of high-cost, long-lead-time items to items of successively lower cost and shorter lead times until all desired items were covered. Greatly increased data processing capability has permitted MRP application to myriad items.

Present-day MRP programs must meet five data prerequisites if they are to function:

1. A time-phased master production schedule (MPS) stating how many of each end item is to be produced, and when
2. A bill of material (BoM) for each MPS item and for each parent item (any item having one or more components) at lower BoM levels which MRP is to plan
3. A unique number for every MPS item and all parents and components in BoM
4. Files of inventory data for all stocked items
5. Lead times for every purchased and manufactured item

These so-called "must have" data are mandatory. In addition, other "nice to know" data, covered in detail in Chapter 7, are useful enhancements to MRP. These include part item master data (allocated quantities, order quantity modifiers, scrap factors, ABC codes, responsible planner identification codes, year-to-date usage, and others), BoM data (initial release dates, engineering change numbers and effectivity dates, family codes, "run with" codes grouping orders for setup or processing reasons, responsible-engineer codes, and others), and MPS data (sales forecasts, actual orders, critical work centers, product family codes, and others). In addition, MRP needs support from other subsystems, including:

1. Customer order processing
2. Process information
3. Inventory transaction handling
4. Open purchase order tracking
5. Open manufacturing order tracking
6. Production reporting

The only language MRP understands is numbers and letters: MPS items, inventoried items, ordered items, quantities in stock and on order, lead times, and other numerical and alphabetical data. MRP cannot work with English-language product descriptions, sales catalog model numbers, or ambiguous numbers that fail to identify precisely the items to which they apply.

Part Numbers

Each inventory item must be identified by a part number. The purpose of the number is simply *to provide a unique name* for each individual, as people's names do. Parts having any differences in form, fit, or function, so that they are not interchangeable, must have different numbers. Ideal numbers should have the fewest digits, be only numeric (no alphabetic characters), and be assigned serially as new parts are introduced.

A trap into which many fall, greatly complicating the problem of keeping accurate records, is doing more with part numbers than giving each a name. Digits at each position are given significance, identifying some characteristic such as shape, material, or product family. This lengthens the number, shortens its useful life, and increases the likelihood of people making errors writing it or keying it into computer-based systems. Significant-digit numbers are a holdover from punched card data processing, when the number of columns available for item data was limited.

Proponents of significant digits argue that data processing equipment handles longer numbers at negligible cost. They overlook the work of adding more new numbers (features change much more frequently than form, fit, or function), correcting more errors, and needing even longer numbers (a single digit can handle only 10 varieties of a characteristic). They also forget that computers can access subordinate files of code numbers for descriptive data without needing these codes in part numbers.

It is difficult to argue against semisignificant numbers where only two or three digits identify families of parts. Those desiring such numbers should be responsible for justifying them by showing that advantages outweigh disadvantages. It is rarely possible to replace existing part numbers in a going concern; costs and other demands on scarce resources are prohibitive. The simple solution is to adopt short, serial, nonsignificant numbers for all new items, deleting old ones as they become obsolete. Changes now proliferating in most businesses will lead them to make the conversion in a remarkably short time.

Bills of Material

Each item in the MPS must have a unique identifying number and be associated with a BoM specifying its components to be planned and controlled by MRP. Bills of material identify the component(s) needed to make parent items. A parent may be as complex as a product assembled from many components or as simple as a single part made from some raw material.

Product structure data can be stored in computers using BoM processing software commonly supplied in computer manufacturers' and good commercial software. These utilize computer storage efficiently, avoid duplication of data, and apply fast retrieval for "assembly" by the computer of BoM in various formats desired by different users. Bills can be printed out or displayed on CRT screens as desired.

Unfortunately, in practice, some parent items may have as many as five different bills of material:

1. *A design parts list.* Simply listing the component items, this BoM is the last step in engineering design. It is the engineers' way of telling the rest of the organization how many of which components make up a product. Parts lists may show how engineering thinks the parent should be put together, but this is usually not the design engineers' responsibility. Engineering specifications accompany BoM and carry other information needed to produce, inspect, and test complete, functioning products. Parts lists often do not include packaging materials and usually omit items like glue, grease, and paint for which it is difficult to specify a quantity needed.

2. *A manufacturing BoM.* In addition to listing all components of a product, BoM must be structured to show production people how to put them together. Production may require subassemblies for proper welding or ease of assembly. Semifinished parts (unpainted, not plated, incomplete machining) may reduce complexity and increase planning and production flexibility. Sales of field replacement spare parts may be assemblies made only for this purpose. Engineering usually has no interest in these needs.

3. *A material planning BoM.* Development of valid, realistic MPS for products that offer several options to customers require very different BoM from those issued by engineering and those needed by production. Material planning for many varieties of the same basic product, for tooling and similar related materials, and for make-to-order products made from a few standard subassemblies require specially structured BoM. This subject is covered in Chapter 7.

4. *A cost-accounting BoM,* often simplified by using one part number for many painted or plated parts and for other components with variations not affecting costs or inventory valuation.

5. *A BoM describing the actual item made,* which for several reasons is different from all other BoM for the item.

There is only one legitimate reason why manufacturing BoM should differ from the way products are actually built: last-minute design changes issued by engineering as they were being built were not yet picked up in computer files. Strenuous efforts should be made to keep such time delays to a minimum.

BoM showing products actually made are often different from planning BoM. This can be caused by legitimate differences; what was planned was not what was built. Too often, however, the reason these BoM types are different is lack of data accuracy in the formal files. Each group using BoM attempts to keep its own; inevitably differences creep in.

The five types of BoM are all needed. This does not mean that five different BoM computer files are necessary; bill-of-material processor programs can code the basic data to link components and produce BoM for each specific purpose.

The term *bill of material* is used interchangeably for that covering a single parent and its components, called a *single-level BoM,* for more complex *multilevel BoM* having several levels, and for the entire bill-of-material computer file.

Figure 3-1 depicts the product structure of an item A-10001 in a single, composite bill of material, usually called a "tree," showing all levels. Early manual methods of storing product-structure data used this format because it minimized the time needed to look up BoM for each subassembly. Obviously, where components are common to several parents, much duplication exists, and maintaining BoM with frequent engineering changes is difficult.

When only manual BoM were used, it was easier to take a BoM for a product or model and create one for a new model by simply adding new parts and deleting old ones not used. This *add-and-delete* type of BoM, also called *comparative* and *same-as-except,* minimized paperwork for design engineers and highlighted component differences for planning, production, and cost-accounting people. They are awkward for use in computer-based MRP programs, however, and only rarely used.

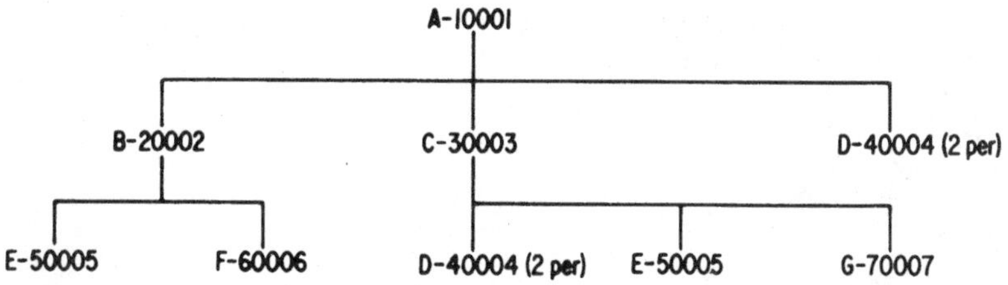

Figure 3-1. Product structure, Item A 10001.

For efficient storage and easy maintenance, only single-level bills are loaded into computer files. A single product-structure record for each parent item is established; it simply lists the parent item's next-level components, showing identity, quantity per assembly, an address, a code identifying component items that are also subassemblies, and another specifying user interests. In the last case, for example, "E" would identify design engineering, "P" planning, and "M" manufacturing; these enable printing or displaying BoM differently for specific users. Multilevel BoM are reconstructed by BoM processing programs "chaining" or linking addresses of components.

Product-structure data of item A-10001, shown in Fig. 3-1, would be stored in single-level bill format as follows:

Parent:	A-10001	B-20002	C-30003
Components:	B-20002*	E-50005	D-40004 (2 per)
	C-30003*	F-60006	E-50005
	D-40004 (2 per)		G-70007

*Subassemblies

When not otherwise indicated, quantities per assembly are one. In this example there are three single-level bills, A-10001, B-20002, and C-30003. The data can be formatted and displayed in various ways; the six most popular formats are:

1. Single-level explosion
2. Indented explosion
3. Summarized explosion
4. Single-level implosion
5. Indented implosion
6. Summarized implosion

The *single-level explosion* format is a single-level BoM. For item A-10001, the following data would be displayed:

A-10001
B-20002*
C-30003*
D-40004 (2 per)

The *indented explosion* format lists components on all lower levels, indenting component numbers under their respective parent's number. Indentations signify levels. By convention in the United States, BoM levels are numbered from top to bottom; end items in MPS are usually 0, although occasionally numbered 1. For item A-10001, the printed output, also known as an *indented parts list,* would appear as

A-10001

- B-20002
  - E-50005
  - F-60006
- C-30003
  - D-40004 (2 per)
  - E-50005
  - G-70007
- D-40004 (2 per)

The *summarized explosion* format lists all components of all parents regardless of level; quantities per assembly reflect total use per unit of the top-level parent. For item A-10001, the printed output, also known as a *parts list,* would appear as

A-10001

B-20002
C-30003
D-40004 (4 per)
E-50005 (2 per)
F-60006
G-70007

The *single-level implosion* format, commonly called a *where-used list,* shows all parents of a component one level up. For item E-50005 these are

E-50005

B-20002
C-30003

The *indented implosion* format traces component usage upward to its parent, then to its parent's parent (grandparent), until the top-level end item is reached. Even the simplest products have at least two levels. The implosion for item F-60006 would appear as

Looking downward	or	Looking upward
A-10001		F-60006
B-20002		B-20002
F-60006		A-10001

The *summarized implosion* format is an expanded where-used display listing all parents on higher levels that contain a particular component. The quantities per assembly are total quantities of the item used in all end items. For E-50005, the implosion would be

E-50005
A-10001 (2 per)
B-20002 (1 per)
C-30003 (1 per)

Other BoM formats are useful in MRP programs. Inverted BoM help plan packaging materials. Figure 3-2 compares the conventional BoM format for film packs to an inverted one. The packaging materials in the latter are shown as components of the film rather than the film pack. The bulk film, easier to forecast, becomes the MPS end item rather than the several pack varieties. Fractions provide the quantity per assembly (Q/A) data for packaging materials, and these are overplanned (total = 1.1) to plan extra sets of inexpensive items.

Tight control of expensive tooling (grinding wheels, form cutters, and chemicals) "consumed" in processing the materials used directly in products can result from adding these items to BoM as components. MRP, as it does for all items, will determine requirements, net available supplies, and schedule replenishment orders. Careful estimates (or several revisions) are needed to get the right quantity per parent item, and units of measure should be chosen to avoid very small decimal fractions.

Other uses for BoM are preparing material requisitions or storeroom picking lists, showing how assemblies go together, product costing, handling engineering changes, and helping solve problems. To meet specialized needs, additional display formats can be created via special user-written retrieval programs.

Maintaining BoM accuracy is neither simple nor easy. BoM frequently have too many levels, adding excess parent numbers, and often responsi-

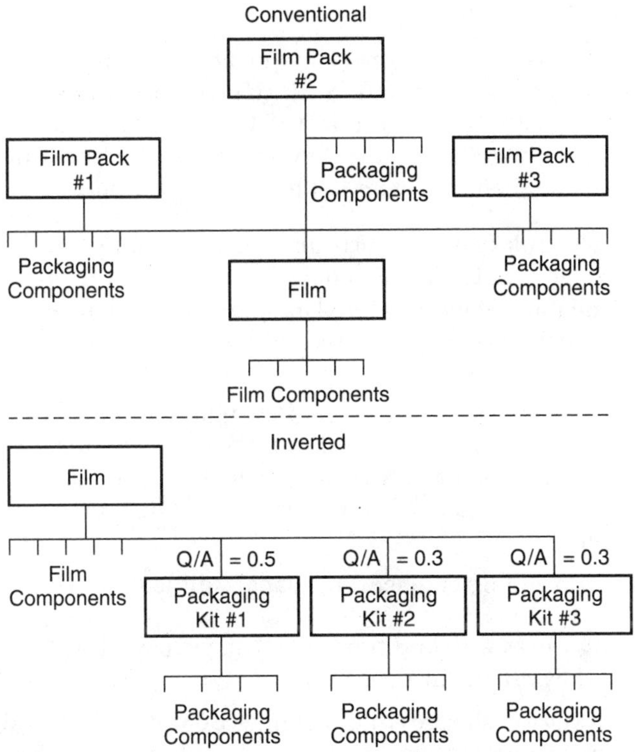

Figure 3-2. Conventional and inverted BoM.

bility for BoM accuracy is not clearly assigned. Keeping them accurate is complicated by frequent engineering changes, some mandated as "immediate," some optional to be phased in later; timing such changes properly is almost impossible with long lead times and poor storeroom and plant discipline. The next section covers control of engineering changes specifically.

The bill-of-material file (or product-structure file) guides the MRP explosion process, as is described in Chapter 4. In this, product-structure data are not operated on but merely consulted by the BoM processor program to determine component identities, quantities per assembly, and addresses.

Chapter 7 contains an additional discussion of BoM structuring to improve master scheduling, particularly for products with optional features for customer selection.

Engineering Change Control

The central role BoM play in MRP programs makes it essential that engineering changes be handled correctly. Such changes are introduced into BoM by "effectivity dates" entered into BoM process file data for individual items affected, together with references to the engineering change numbers initiating them. Changes fall into two broad categories:

1. *Mandatory.* These result from product functional failures, potential harm or injury to users, unavailability of materials, and new government laws or regulations, and must be implemented immediately, perhaps even recalling products in customers' hands. Costs and internal problems are irrelevant.

2. *Optional.* These include the great bulk of changes resulting from competitors' innovations, new materials and processes, cost reductions, and improving operations. Timing is usually specified as "Soon as possible," When current material gone," "Minimum cost," "On X date," or "With Y serial number."

Four factors make controlling engineering changes difficult:

1. Design engineering works with and transmits data via parts lists that have been modified by others for their use.
2. The timing of introducing changes into planning BoM is critical, and many factors deserve consideration.
3. The sheer volume of changes can be daunting.
4. The effects on all BoM levels can be difficult to trace.

Engineering parts lists usually do not contain all items needed by others using product structure data, as is explained earlier in this chapter. Therefore, all effects of engineering changes are not apparent to design people. Locating the specific items covered by changes is not always easy in restructured BoM. Engineers often are reluctant to issue new part numbers if they perceive that form, fit, and function are unchanged and parts will be interchangeable but existing numbers may not provide the unique part numbers that MRP needs to function.

MRP uses BoM as a framework for planning and replanning; the *timing of changes* affects the ways it handles requirements, netting, and order scheduling. Factors to be considered include

1. Competitive advantages
2. Legal liability

3. Cost benefits
4. Inventories of current materials
5. Availability of new materials
6. Service part needs
7. Tooling and other equipment
8. Effects on machine and equipment capacity
9. Customer order groupings and similarities
10. Documentation (parts lists, instruction manuals)

Adequate consideration of these requires input from many sources, involving many people with different perceptions. A team approach is commonly used to do this job well.

The number of engineering changes released in any period always amazes the people who have to handle them. Changes can be caused by failure of products to function properly, customer preferences, competitors' innovations, materials and process improvements, cost reductions, and many other factors. Changes rarely come uniformly; floods follow immediately after new or radically revised product designs have been released. Teamwork during application design, and sometimes even in research and development, reduces greatly the number of later engineering changes. Called *concurrent engineering,* the teams usually include design and processing engineers, sales and marketing people, planners, quality, production, and cost-accounting people. A more detailed discussion of concurrent engineering is found in Chapter 12.

Groups of engineering changes for periodic release, called *block changes,* have significant benefits over individual releases. These include more stable BoM, more accurate maintenance at lower costs, minimal disturbance of operations, simpler service parts support, reduced obsolescence, and better planning.

The complexity of product structures in many firms, coupled with restructuring, often *obscure the full effects of apparently simple engineering changes* on both lower and higher BoM levels. People evaluating these effects must understand the reason for the changes as well as the mechanics of BoM structuring.

Tight management of engineering changes requires

1. *Clear assignment of responsibility for BoM accuracy.* This is often given to one group, often called Data Management, who coordinate the work of all groups involved and make final decisions on BoM data integrity. They also audit BoM files to detect errors and identify causes.

2. *Easy accessibility of BoM files.* Used by nearly everyone in manufacturing businesses, BoM data must be available easily and quickly. BoM processor programs handle requests, but computers must be running when needed and open for inquiry.

3. *Clear audit trails, called* configuration control, *to trace the history of BoM from first introduction to final obsolescence.* Defense and aerospace firms, and those dealing with food, medicines, and radioactive materials, have legal and contractual obligations for these. The possibility of violating governmental regulations and incurring large penalties for injury or death is making more companies maintain these audit trails.

Master Production Schedules

This is one of the most vital sets of data in manufacturing planning and control. MPS are managements' "handle on the business" by which they authorize all actions of their people involved in manufacturing products and serving customers. MPS are planning devices, statements of what can and should be made, means of balancing customers' needs against plant capabilities, and bases for coordinating all organization groups and measuring their performance. *They are not execution tools, sales forecasts, customers' orders, or final assembly schedules.*

Does every manufacturing company or plant have MPS? If they are defined as overall plans of production, it would be difficult to conceive of a plant operating without them. Some manufacturing managers say that they do not have master production schedules; they really mean that their overall plan of production is not expressed in one formal set of numbers. In any manufacturing operation, the sum total of what a plant is committed to produce at any point in time is equivalent to MPS totals. For MRP, the creation and maintenance of formal MPS are prerequisites.

MPS should not be confused with forecasts. These represent estimates of demand from outside sources, whereas MPS constitute plans of internal production. These are not necessarily the same; products are often made at different times and in different quantities from those desired by customers. A clear distinction exists between developing forecasts and setting up schedules of production, despite the fact that in some cases the two may be identical in content. Forecasting attempts to predict when products will leave inventory; final assembly scheduling, supported by MPS, plans when they will enter.

MPS are statements of how many specific items will be made and when they will be made. As covered in Chapter 8, "items" in MPS are end items, defined as highest-level entities not a component of any parent in BoM used by MRP for exploding requirements. End items may be products, major assemblies, planning modules, groups of components covered by pseudobills, or even individual parts used at the highest level in the product structure or subject to demand from sources external to the plant.

Forecasts of all external demands are technically part of MPS but may not be listed in the formal MPS files; some for components are entered as gross requirements in their inventory records. Real orders are inputs to the execution phase and should not be entered into MPS. The reasons for this and the mechanics of handling such orders are covered in Chapter 10.

The MPS format normally is a matrix listing end items vertically and quantities of each horizontally in periods. Figure 3-3 is typical of MPS for families of end items having numerous models. The meaning of the quantities in relation to the timing indicated is fixed by users' convention. They may represent end-item availability, the start of end-item production, or end-item components availability. Depending on which is adopted, the interface between the master production schedule and the MRP system will vary; this is discussed in Chapter 4.

Time periods of MPS must be identical to those of MRP; typically they are one-week periods. Sales forecasts and production plans used by management and marketing, however, are developed and stated in months or quarters, often for generic models. These data must be translated into weekly figures for specific end-item numbers. Figure 3-4 shows this relationship.

The time the MPS spans, termed the *planning horizon,* is sometimes divided into firm, flexible, and unrestricted portions as indicated in Fig. 3-4; the dividing lines are called *fences.* The firm portion, equal to the cumulative (procurement and manufacturing) lead time, is viewed as firm (not "frozen"), representing quantities of end items for which component materials have been procured and work begun on some of them.

	52 Weeks								
Item A	50	–	50	–		60	–	60	–
B	610	610	610	610		610	610	610	610
C	340	360	380	400		300	310	320	330
D	180	180	180	180		150	150	150	150
E	15	–	–	–		–	–	20	–
F	205	205	205	205		225	225	225	225
Family Total/Mo	5575					5340			

Figure 3-3. Typical MPS format.

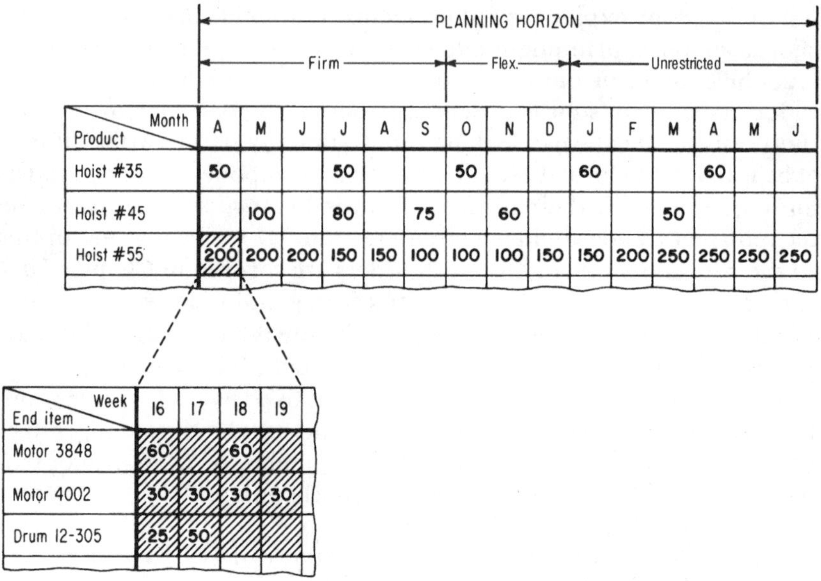

Product \ Month	A	M	J	J	A	S	O	N	D	J	F	M	A	M	J
Hoist #35	50			50			50			60			60		
Hoist #45		100		80		75		60				50			
Hoist #55	200	200	200	150	150	100	100	100	150	150	200	250	250	250	250

End item \ Week	16	17	18	19
Motor 3848	60		60	
Motor 4002	30	30	30	30
Drum 12-305	25	50		

Figure 3-4. MPS horizon segments.

The frequency of MPS maintenance (updating, revision) is usually geared to the forecasting cycle. This is often weekly when computer forecasting models are used, but may be monthly when management reviews the data. Between "official" updating, however, MPS revisions may be needed to respond to significant changes in the mix of new customer orders and to serious unplanned developments in procurement and manufacturing.

While it is desirable for MRP to be capable of responding to unofficial changes in MPS frequently, this must be balanced against "nervousness," which produces myriad notices of changes in order schedules. Arguments in favor of and against daily, or even continuous, replanning are presented in Chapter 10.

Lot Sizes

The particular technique used to determine order quantities for a given inventory item also determines the requirements for its components. In carrying out a complete MRP explosion, lot-sizing techniques and codes selecting the proper one for each item are needed in the computer program that controls the requirements computation. Lot sizing is covered in Chapter 6.

Data Integrity

A precondition for effective operation of MRP programs is a high level of file data integrity. File data must be accurate, complete, and up-to-date, if MRP is to reach its full potential. MRP can function with faulty data and generate some useful outputs; there are good reasons (debug programs, train people, measure benefits, and cope with a crisis) to start applying MRP before all records are as accurate as desired. If this is done, however, top priority must be given to finding and fixing the causes of errors quickly before the MRP's credibility is destroyed.

The requirement of file data integrity may seem self-evident. To be effective, planning systems must provide credible data; errors destroy credibility. Order point techniques do not use MPS and BoM, and the effects of errors in inventory data are minimized by safety stock cushions; their users become resigned to acceptance of the need for expediting.

Order point acts merely as an order-launching system, a "push" system that must be complemented by an expediting "pull" system in order to function at all. After orders have been launched, order points serve no function; they cannot be replanned. The true need dates for orders at subsequent operations, in stockrooms, or at assembly lines must then be determined by the "informal system" using shortages or other priorities.

An MRP program is capable of providing both push and pull functions; there is no need for the informal system *if the data it uses are correct.* Integrity of MPS, BoM, and inventory files in particular is vital to MRP. The meticulous maintenance of these files calls for dedicated, special efforts by system users—something easy to understand, but very difficult to get.

Common assumptions about data integrity were proved wrong many years ago. It is not expensive, it is not time-consuming, and it is not impossible to get accurate records. It does require understanding the prime objective, not to find and fix errors, but *to find and eliminate the causes of errors.* This takes a coordinated attack by all those handling data, each assuming responsibility for the quality of data handled.

MRP Assumptions

Several assumptions, some explicit and some implicit, are made in MRP-based planning and control programs:

1. Every inventory item moves into and out of stock.
2. All components of an assembly are needed at the time an assembly order is released.
3. Components are disbursed and used in discrete lots.
4. Each manufactured item can be processed independently of any other.

MRP assumes that *every inventory item under its control passes into and out of stock,* and that there will be reportable receipts, following which the item will be (even if only momentarily) "on-hand." Later it will be disbursed to some "customer"—either a production operation, warehouse, or external customer. This dictates the need for tracking and recording the movement (physically or figuratively) of items into and out of storerooms.

Standard MRP logic assumes that *all components of an assembly must be available at the time an order for that assembly is to be released to the factory.* This is the time associated with each component's gross requirements. It is assumed also that unit assembly lead time (the time required to produce one unit of the assembly) is relatively short and that all components are needed simultaneously. For most assemblies, this assumption is true.

In cases of significant exceptions to this rule, where it may take several weeks to assemble a unit and expensive components are needed successively over this period, standard MRP procedure can be modified by setting up sub-BoM groupings of components needed early and those needed later during assembly.

Another assumption under material requirements planning is *discrete disbursement and usage of components.* For example, if 50 units of a component are required for a given work or assembly order, MRP expects that exactly 50 units will be disbursed and used. Bar stock, continuous sheets or coils, or large liquid containers are usually issued in unit quantities, often larger than needed; this requires that standard MRP programs be modified to handle such inventory items properly.

The problem is recognizing that excess material issued is still "available" for use in other orders but is no longer in the normal stock area. One solution is to have the program calculate the excess of issued over required quantity and to add this to a field in the item's inventory record called "excess issue," which is relieved first when later orders calling for it are released. For normal future requirements planning, the sum of this and the "on-hand" total are added. This greatly complicates cycle counting to verify inventory record accuracy, of course.

Another assumption implied by MRP is process independence; *any manufacturing order for any inventory item can be started and completed independently of any other order.* A "mating part" relationship (where one item at its third operation must be joined by another at its fifth operation to machine a common surface) and setup dependencies (where orders for several items should be set up in sequence) are exceptions and need program modifications. The common solution to this problem is to include "run with" codes in each items' records, alerting planners to schedule order releases together. Clever applications of lot-sizing techniques will ensure that items in such families come up for ordering at about the same time.

Essential Data

All MRP systems have a common objective: to provide necessary data to planners to aid in managing inventories and production. MRP translates a time-phased master production schedule of end items into component gross requirements, showing them as discrete period-demands for each item planned. The MRP "netting" process follows calculation of gross requirements, applying existing on-hand inventories and released orders to meet at least part of these gross requirements.

MRP next develops order recommendations by computing net requirements for each inventory item, time-phasing them, and determining the proper timing of order coverage. Planned orders cover both procurement (purchase orders) and production (shop orders). The mechanics of all of these calculations are described in detail in Chapter 4. The essential data utilized in planning new order actions are

1. Item identity number (part number)
2. Order quantity recommended (lot size)
3. Date of recommended order release (start date)
4. Date of indicated order completion (due date)

New order action for purchased items usually takes two steps: sending requisitions to purchasing from inventory control, and, subsequently, placing purchase orders with suppliers. More recently, inventory planners, called *planner-buyers,* place individual purchase orders (actually, releases against blanket-type family orders placed by purchasing) directly with suppliers.

In addition to new order actions, MRP recommends revisions to previously released order actions, limited to

1. Increasing order quantities
2. Decreasing order quantities
3. Canceling orders
4. Advancing order due dates
5. Deferring order due dates
6. Suspending orders (indefinite deferment)

MRP is capable of planning for safety stock, although it is included in rare cases only. The calculation of net requirements with safety stock views it as another requirement. It can be handled in two ways: added to net requirements, or prededucted from on-hand inventory before MRP begins time-phasing gross requirements. The latter is the common way.

MRP time-phases gross and net requirements. Net requirements are covered by planned orders; order quantities either match net requirements or are calculated using one of the several lot-sizing techniques described in Chapter 6. The timing of planned order releases is determined by MRP, and the information is stored or printed out for future order action.

Coverage of net requirements is achieved only in part through planned future orders. Since the netting process includes "on-order" amounts, the MRP system also reevaluates the timing of released orders in the near future and signals the need for rescheduling these orders, forward or backward in time, if required for coverage of net requirements. The mechanics of this are described in detail in Chapter 4.

The preceding discussion of prerequisites and assumptions raises the question of MRP applicability to different types of manufacturing businesses. If some prerequisites cannot be provided or if some assumptions are invalid, can MRP be used effectively? MRP represents sound planning practices valid in all types of manufacturing. Inventory items should and can be uniquely identified, a bill of material should and can be created, integrity of file data should and can be maintained. Some exceptions can be handled by changes to standard MRP programs as described above for fictitious stocking, spaced delivery of components during long assemblies, and excess issues.

It is true that the first companies to develop and use MRP were manufacturers of complex assembled products, involving procurement and fabrication of many components with large numbers of orders in production simultaneously. This environment represents the most severe inventory management and production planning problems. To alleviate these problems, companies reached for MRP as soon as computers made it feasible.

It is clear that the application of MRP is generally better in discrete order as contrasted to continuous processing and repetitive manufacturing, where specific batch orders may not exist, and for complex rather than simple, even one-piece, products. Even in such operations, however, MRP can be applied effectively using the modifications described in Chapter 12.

Since the pioneer days of MRP, it has been used successfully in so many diverse businesses (cable and wire, furniture and packaged spices, for example) that applicability criteria have been obscured. The principal criterion of MRP applicability is the existence of MPS to which raw material procurement, fabrication, and subassembly activities are geared. This criterion is met in practically all manufacturing. MRP is primarily a component procurement/fabrication planning technique.

The important criterion in deciding whether purchased or manufactured items in a plant using MRP should be brought under it is the type of demand, independent or dependent; this is discussed in Chapter 2. Other item attributes such as cost, volume of usage, or continuity of demand are irrelevant to MRP.

Whether or not low-cost, numerous C-items deserve the full MRP treatment is debatable. MRP arguably processes large volumes of data at very low cost, but this is the least of the costs involved. Skilled people, a scarce resource almost everywhere, must maintain files, review MRP outputs, make decisions, and take action. The potential gains from MRP treatment of low-value items are small; they can easily be outweighed by better control of high-value A-items and medium-value B-items. Simpler techniques for C-items, described in Chapter 2, can handle them well outside MRP. This is particularly true of items with common usage, those quickly and inexpensively purchased or produced, and inexpensive, rugged items with long shelf life.

Lead Times

MRP requires procurement and production lead-time data for all inventory items; the elements making up these lead times are described in Chapter 8. Planned lead times are supplied by planners and normally have fixed values, but these can be changed at will. The potentially harmful effects of the nervousness generated by frequent changes in planned lead times are pointed out in Chapter 10. More than one value cannot be used simultaneously for an item, but a program modification can be made to alert planners to orders coming up for release in some predetermined time period. This time can be used for paperwork or other preparation for release without adding to planned lead times.

Lead times of individual items are used in calculating the timing of release dates from scheduled completion dates for planned orders. This offsetting process is described in Chapter 4. MRP assumes that a component-item order must be completed, at the latest, in the period containing the parent-item requirement that needs it. Given an item lead time shorter than its time period (bucket), MRP will act as if it were zero, calling for starting and completing orders in the same period.

How manufacturing lead times relate order start and finish dates is best illustrated in an example. Lead times for four items with parent-component relationships are

Transmission *A*:	1 week
Gearbox *B*:	2 weeks
Gear *C*:	6 weeks
Forging Blank *D*:	3 weeks
Total:	12 weeks (cumulative lead time)

Transmission *A*, used in assembling Truck *X*, contains 1 Gearbox *B*, which, in turn, requires 1 Gear *C*, made from 1 Forging Blank *D*. If assembly of Truck *X* were scheduled to start week 50, order release and completion dates would be calculated by successively subtracting the component lead times from 50. If Transmission *A* were needed in week 50, Gearbox *B* would have to be available in week 49, and Gear *C* in week 47. Forging Blank *D*, needed by week 41, would have to be ordered in week 38.

Release dates of parent orders establish the timing of gross requirements for their component items, as illustrated for the gear and forging in Fig. 3-5. If on-hand inventory is less than this gross requirement, the component will have a net requirement in the same period (4). The completion date of the order covering this net requirement will be geared to this timing; its release dates will be set by subtracting the component's lead time. Positioning the planned-order release forward of the timing of the net requirement it covers is called *offsetting for lead time*. This topic is covered fully in Chapter 4. This example assumes that lead times will not vary with the order quantity.

Planned, sometimes called *normal* lead times, should be a reasonable average of actual lead times, unless a decision has been made to reduce them significantly. In normal MRP operations, their accuracy is not crucial; planned lead times are used merely to determine order release dates, which are considerably less important than completion dates. Other considerations affecting the validity of priorities, covered in Chapter 9, and in improving operations, covered in Chapter 11, are even more important.

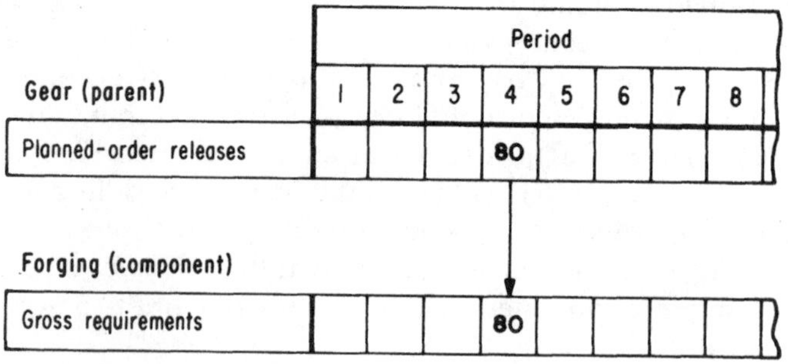

Figure 3-5. Timing of a gross requirement.

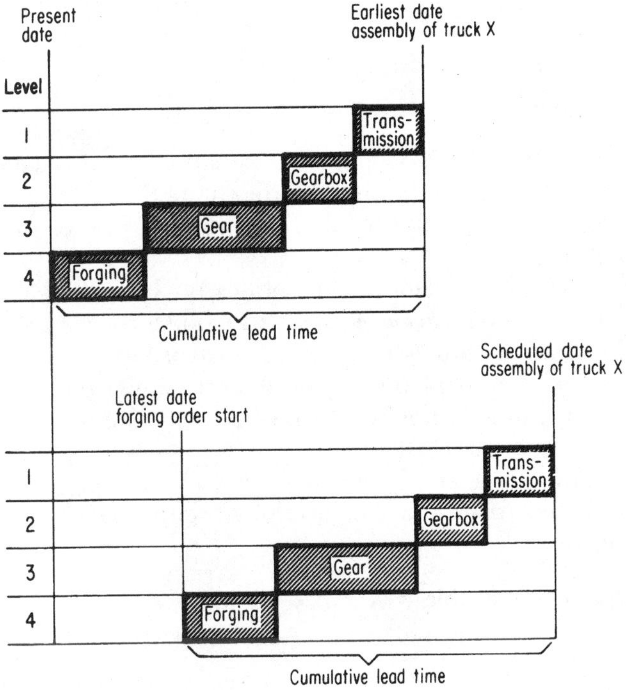

Figure 3-6. Cumulative lead time.

The sequence of lead times of the four items in the example is called the *cumulative lead time,* graphically represented in Fig. 3-6. If lead times cannot be overlapped, this can be viewed as the critical path that determines the earliest time that the end products could be built or, given the end-product schedule date, the latest time for the start of the lowest-level item order. Ways are available to collapse lead times (lap-phasing, order-splitting, multiple machines, and others) but normally these are not planned; they are used to meet dates when orders are urgently needed.

Inventory Status

MRP requires inventory status records for all items under its control, including quantities in stock, on order, allocated, and available. The inventory or stock status of an item must be known before MRP can determine what, if any, action is to be taken on that item. This is expressed by data showing an item's current situation. Status information provides answers to the essential questions of

1. What do we have now?
2. What do we need in which periods?
3. What must we do to meet these requirements?

The answer to the third question is obtained by evaluation of status data by an inventory planner. Computers can be programmed to perform evaluation procedures routinely, but this should be done carefully and only after long experience.

The simplest inventory status data are quantities on hand and on released orders; their sum is called *available*. Needed action is determined by comparing requirements to status, scheduling planned orders when the available total is zero, or upon depletion of this total to some predetermined minimum, at which time action will be taken to meet anticipated future needs.

Better inventory status data are provided by the classic perpetual inventory control equation that sums up inventory status, albeit still without time-phasing:

$$\text{Quantity available} = X = A + B - C$$

where A = quantity on hand
B = quantity on order
C = quantity required

If X is positive, there is some quantity available for future requirements; if it is negative, an impending shortage may occur because of inadequate coverage. The objective of sound inventory control, of course, is to keep X equal to or above zero at all times. MRP attempts to do this by placing a new order (increasing the value of B) whenever X approaches zero or goes negative. While this policy would appear to preclude shortages, it does not; the status data are inadequate in three ways:

1. They lack information on timing.
2. The data on B and C may represent totals of several separate inventories and orders.
3. The status formula does not provide for planned (future) coverage.

For example, status might be as follows:

On hand = 100 On order = 120 Required = 200

$$X = 100 + 120 - 200 = 20$$

It appears that all is well and no action is required now; in fact there will be a shortage, as becomes evident when timing data are added:

On hand = 100   On order = 120 due June 1

Required = 200 needed by May 15

The quantity is adequate, but the timing is too late. To illustrate the opposite case, if the status were

On hand = 20   On order = 100   Required = 200

$$X = 20 + 100 - 200 = -80$$

it would seem that an additional order for 80 or more pieces should be released now, March 1st. For order lead time equal to four weeks, it would be scheduled for completion on March 29. This would be wrong, however, as the planner would see immediately if he had these timing data:

On hand = 20   On order = 100 due March 10

Required = 110 needed March 15 and 90 needed June 15

All that are really needed are 80 more on June 15. Depending on lot sizing, 80 or more may be run on the additional order.

The simple inventory status equation does not make the true needs sufficiently clear. In both examples, the action signals are false. The correct action message in the first case should have been, "Advance the open order from June 1 to May 15" and in the second, "Do nothing now, but an order for at least 80 will have to be released around May 15." The equation uses only released orders and cannot indicate a schedule of planned orders.

MRP overcomes the above shortcomings by adding planned order data and by time-phasing all status data. The classic equation, in effect, becomes

$$\text{Available} = X = A + B + D - C$$

where *D* equals a quantity planned for future order release.

Program Design Features

Much of the design or architecture of MRP programs has been standardized in a multitude of application software available from computer manufacturers, software houses, and consulting firms. Most working MRP programs utilize standard software comprised of optional modules for the major elements; only a small fraction have been designed and programmed by users. Very few MRP programs are identical, however; users have strong (if erroneous) convictions about their specific needs in item transfer data formats, report contents, and decision-making subprograms.

Standard MRP software provides users with considerable freedom in "customizing," providing a selection of modules for master production scheduling, bill-of-material processing, lot sizing, purchase and manufacturing order handling, and other functions. The pendulum of complexity has swung far beyond the point of diminishing returns; it is now clear that companies in the same type of business (these types include batch, process, continuous, make-to-order, and even aerospace/defense) can make effective use of one MRP program configuration. What justifies different ways of processing bills of material, calculating gross and net requirements, offsetting lead times, and tracking released orders?

Practically all suppliers of MRP programs provide a comprehensive core-system package for manufacturing planning and control, including MRP, rough-cut and detailed capacity planning, and capacity and shop floor control, plus many support programs for procurement, design and process engineering, quality management, cost accounting, and plant maintenance. Unfortunately these are usually called "MRPII systems," leading to confusion between the material requirements planning technique and the manufacturing resource planning system they represent. Sophistication of programming of these systems is also overdone.

The effectiveness of planning and control depends very little on the elegance of computer programs. The system is a C-item in the total picture; its ability to help its users run their businesses well depends on *how well they use it, not on how well it has been designed technically.*

The three principal functions of MRP are to

1. Plan and control inventories
2. Plan and replan released order priorities
3. Provide data for capacity requirements planning

Design of MRP programs should meet these objectives:

1. Planning inventories
    a. Order the right part
    b. Order in the right quantity
    c. Order at the right time
2. Planning priorities
    a. Order with the right due date
    b. Keep the due date valid
3. Planning capacity
    a. A complete load
    b. An accurate (valid) load

c. An adequate time span for visibility of future load

Meeting these requirements requires making the proper design decisions for

1. The span of the planning horizon
2. The size of the time-bucket
3. The coverage of inventory by class
4. The frequency of replanning
5. Traceability using pegged requirements
6. Capability of "freezing" planned orders

Planning Horizon

To ensure that MRP provides data on items at all levels in bills of material, the planning horizon should at least equal the largest total of item lead times (cumulative lead time) in the critical (longest) path leading from raw material to the end item appearing in the master production schedule. If planning horizons are too short, the process of successively offsetting for lead time in level-by-level planning will run into past periods when it reaches items on the lowest level. To ensure some "forward visibility" of data on purchased items, planning horizons should be significantly longer than the critical path lead time.

This is illustrated in Fig. 3-7, where the cumulative lead time is 15

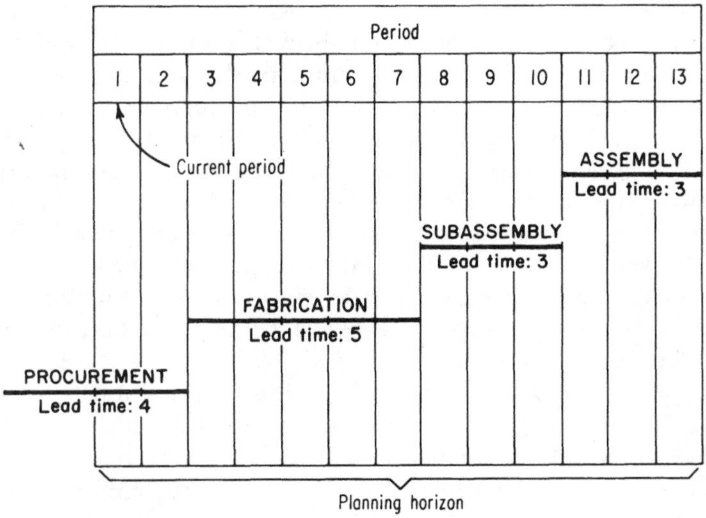

Figure 3-7. Planning horizon and cumulative lead time.

periods but the planning horizon only 13 periods. The order for purchased material, developed by the system through the explosion of an end item at the end of the planning horizon, should have been released two periods ago, putting it two periods behind schedule before it is even released.

With multilevel product structures, because of successive lead-time offsetting there is a partial loss of horizon at each lower level. The effective planning horizon at each level is successively diminished as MRP progresses from one level to the next. For example, in Fig. 3-7 the fabrication planned order completion is period 7; the effective planning horizon for this item is therefore only 7 periods. Its time-phased inventory record and those of all of its components would show 13 time-buckets, but the last 6 would always have 0 contents.

One consequence of very short horizons is inability to apply some lot-sizing techniques effectively because of lack of sufficient net requirements data. This is discussed in Chapter 6. Another, more serious consequence is lack of data for capacity requirements planning. In the Fig. 3-7 example, a complete load for fabrication operations cannot be projected beyond period 7, although it can be certain that MRP will plan other items using this work center in the future. Short horizons limit capacity requirements planning on the low level, usually fabricated parts, where it is most desirable. However, as is discussed in depth in Chapter 10, long horizons result in plans with low validity because of greater potential changes in requirements, design details, and processing methods, and more upsets. The resolution of this dilemma also is covered in that chapter.

Time-Buckets

When selecting the size of the MRP time-bucket, users face a tradeoff between the desire to have planned events pinpointed in time and the need for clear and simple ordering data. Very small time increments require many MRP program fields and a large number of data elements to be manipulated. Experience has shown that buckets representing one-month periods are too coarse for most users. MRP indicating, for instance, that an order is to be released or completed in October raises the question, "*When* in October?" Given such data, and realizing that shortages have long been viewed as worse than excesses, most planners and buyers will "play it safe"—release orders early and specify early deliveries—thus unnecessarily inflating the level of inventory.

Specifying order completions by month is less than helpful to shop supervisors. They need to know what next to produce this week or, better yet, today. When several hundreds or thousands of shop orders are due in October but MRP schedules provide no relative priority information, shortage lists and expediting will replace the formal system and performance will deteriorate.

Daily time periods are too fine, using too much computer storage and computing time for the results obtained. They give the illusion of precision with dubious accuracy. Planning is not an exact science, in spite of the apparent rigor of MRP calculations. Chapter 1 compares "soft" planning data and "hard" execution data; only the latter need to be precise.

A one-week time-bucket is most common except in a few businesses like food, pharmaceuticals, and fine chemical manufacture, where production times are very short. In other types of manufacturing, one week is reasonable for order releases, completions, priorities, lot sizing, and load reporting.

Replanning Frequency

The frequency of replanning MRP is very important to operating performance. Without replanning, MRP validity gradually deteriorates as requirements and inventory status change. Selecting a replanning frequency, like choosing the time-bucket, is another case of trade-off. Computers love to replan; people like stability. If the environment is very dynamic, more frequent replanning must be considered. Ways must be found, however, to avoid nervous prostration in users of MRP output. Regeneration, net change, and related topics are covered fully in Chapter 5, and coping with the effects of nervousness in Chapter 10.

Pegged Requirements

A special feature called *pegged requirements* provides MRP users with the capability to trace an item's gross requirements to parent sources. MRP calculations progress from top to bottom of the product structure. The gross requirements for a component item, derived from its parent(s) and from additional external sources of demand, if any, are summarized by period. The contents of gross requirements buckets represent total requirements, the sources of which are obscure.

Pegging saves this information at the point in the MRP process when it is known to the program and records it in a special file. Pegged requirements may be thought of as a selective where-used file. A regular where-used file lists all parents of a component item; a pegged-requirements file lists only those parents that have planned orders in the planning horizon that place gross requirements on the subject component. This permits the inventory planner to trace upward level by level in the product structure to determine which parents generated how much of the item's total gross requirements in any period. By following the "pegs" from one item record to another, the planner can trace the demands to their ultimate sources, specific buckets for individual end items in the master production schedule.

An example is provided in Fig. 3-8, in which the demand for item *X* comes from parent items *A, C,* and *D,* and from an interplant or service-part order 38447. MRP programs have fields labeled PARENT RECORD that contain file addresses of parents rather than merely item identification. This facilitates fast retrieval of these records, which is the key to effective use of pegged requirements by busy inventory planners.

The pegging method just described, called *single-level pegging,* requires a succession of inquiries to trace lower-level item demand to the related specific end item in the master production schedule. In bills of material having many levels, this can be very tedious. Full pegging allows planners to link item demand to the MPS with a single inquiry. In this method, each requirement for a component is identified with a specific end-item demand in a time period listed in the MPS.

Full pegging is desirable only in a few situations. It is useful for custom-engineered products, made-to-order products, and standard products having very few or no common components. MRP in most manufacturing environments provides significant benefits by combining parent requirements for commonly used components, setting replenishment order quantities to cover multiple net requirements, and permitting parts on hand and in process to be commingled.

REQUIREMENTS RECORD - ITEM X

Period	1	2	3	4	5	6
Gross requirements	20		35	10		15

PEG RECORD - ITEM X

Period	Quantity	Parent record	External order
1	20	A	
3	15	A	
3	20	C	
4	10		No. 38447
6	15	D	

Figure 3-8. Pegged requirements.

In practice, even single-level pegging is not used frequently. The level-by-level gross-to-net planning process, lot sizing, safety stock, and scrap allowances tend to obscure (or even erase) a clear path through multilevel bills. With today's powerful computers, pegging is feasible for almost every manufacturing company. Like insurance, however, it is nice to have but expensive, and most people hope they never need it.

Firm Planned Orders

A second special capability of MRP is firm planned orders. This technique permits the suppressing of nervousness in MRP caused by the program's freedom to recalculate planned order quantities and reschedule them. Planned orders normally are manipulated by computers using MRP program logic, which determines net requirements and sets each item's planned-order release schedule to cover them using planned lead times and lot-sizing rules.

Replanning revises this schedule automatically when net requirements change, and when dynamic lot-sizing techniques recalculate order quantities. Firm planned orders override computer-driven changes; they are substituted for planned orders when users desire to freeze the quantity and/or timing of planned-order releases. This requires a special MRP program. Figure 3-9 shows how planned, released (scheduled receipt), and firm planned orders are handled in MRP programs.

While computer programs handle changes easily, manufacturing plants and their suppliers cannot always respond, particularly when rescheduling released orders. With firm planned orders, inventory planners can minimize nervousness and the harmful effects of dynamic lot-sizing techniques

	Exploded to Lower Levels?	Rescheduled Automatically?	Exception Messages Generated?	User Control of Quantity, Start & Need Dates?
Planned Order	Yes	Yes	No	No
Scheduled Receipt	No	No	Yes	Yes
Firm Planned Order	Yes	No	Yes	Yes

Figure 3-9. MRP handling of orders.

and avoid excessive numbers of changes in expensive parent item schedules that result in trivial changes in inexpensive components.

They are not unmixed blessings, however. The firm planned order command immobilizes that order in the schedule and prevents MRP from putting another planned order into the "frozen" bucket; this may result in an increased net requirement not being fully covered. This special capability, therefore, should be used judiciously for one specific planned order only, not for an item's whole planned-order release schedule.

Either or both of these two special techniques, pegged requirements and firm planned orders, can be incorporated into basic MRP programs to enhance their usefulness. These features are not essential to the system's operation and are not included in every MRP-software package but they warrant consideration for the specific aids they provide.

PART 2
Methodology

Logic, like whiskey, loses its beneficial effect when taken in too large quantities.

LORD DUNSANY

4
MRP Logic

Logical consequences are the scarecrows of fools and the beacons of wise men.
THOMAS H. HUXLEY

Logic versus Procedure

The term *logic,* when applied to systems, refers to the reasoning behind the procedures and not to specific procedural steps; the validity of the results achieved by procedural steps proves the soundness of that reasoning. Alternative approaches are used in MRP systems installed in different companies, and these involve a spectrum of special features, functions, and procedures. The processing logic of MRP will be described, rather than specific procedures covering all alternative applications.

MRP programs mechanize the fundamental logic of manufacturing as stated in Chapter 2:

1. What will we make? This is stated in modern MRP-based systems in the master production schedule.
2. How many of each component are needed? This is determined from bills of material and quantities in the MPS.
3. How many do we already have? There are two parts to the answer: How many are on hand, and how many already ordered?
4. When do we need the rest? Time-phasing requirements and offsetting processing lead times answer this.

The prerequisites to MRP planning, discussed in detail in Chapter 3, are master production schedules, bills of material, inventory balances, released order balances, and lead times. How these and other data are utilized in the mechanics of MRP must be understood if this powerful technique is to be applied effectively.

Master Production Scheduling

Master production schedules (MPS), defined in Chapter 3, drive MRP, which in turn develops plans for all components with dependent requirements. MPS end items are at the top of the hierarchy of planning bills of material. The link between them and component BoM, called the *master schedule interface,* gives users of MRP three options for choosing what MPS quantities represent:

1. End-item gross requirements
2. End-item production requirements
3. End-item planned orders

The option chosen must be clearly defined and understood by all users if MRP is to function and be used satisfactorily.

Figure 4-1 represents a simple MPS in common matrix format. What does the quantity of 100 *A* in period 1 mean? The three choices are: first, a gross requirement for *A*; second, a production requirement to build 100 *A*; or third, a planned order to be released to build *A*. Its treatment by the MRP system will vary depending upon which option is chosen.

If MPS represents gross requirements, their contents would simply be entered into the gross requirements schedules of the end-items' (*A, B,* and *C*) records and processing would be standard, as shown in Fig. 4-2. The MRP meeting program will deduct on-hand and on-order (Scheduled receipts) quantities, and show that 20 sets of item-*A* components in period 2 and 100 in period 4 need to be available in the first five periods. Under this option, the MPS represents a requirements plan for components needed *to begin end-item production.* Confusion will arise if people view the MPS as a plan of end-item production and MRP does not. This option is rarely chosen.

End item	Period					
	1	2	3	4	5	
A	100		100		100	
B	15	20	25	20	15	
C	50	60		60		

Figure 4-1. A master production schedule.

Item A, Lead time: 1		Period 1	2	3	4	5
Gross requirements		100		100		100
Scheduled receipts		100				
On hand	80	80	80	-20	-20	-120
Planned-order releases			20		100	

20 100

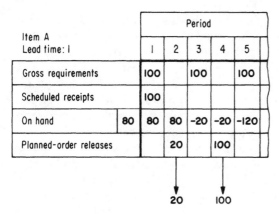

Figure 4-2. Master production schedule interface: gross requirements.

The second choice makes MPS represent end-item production requirements. In the example, 100 units of item *A* are to be finished in each of periods 1, 3, and 5. This presupposes that d*emand for each end item has been netted against its on-hand inventory during the preparation of the MPS.* In this case, MRP must be programmed to exclude end-item on-hand quantities *(but not on-order quantities)* from the netting process, requiring a modification of the regular processing logic applicable to end items only. Figure 4-3 shows this option.

Under each of these two alternatives, MRP will plan to "produce" item *A*, in the quantities and at the times stated in the MPS, and will plan the necessary sets of component parts to support these batches. If the plan is executed well, and not changed, components and other resources needed to make item *A* will be available as scheduled.

Item A, Lead time: 1		Period 1	2	3	4	5
Gross requirements		100		100		100
Scheduled receipts		100				
On hand	80			-100	-100	-200
Planned-order releases			100		100	

100 100

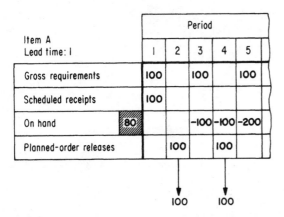

Figure 4-3. Master production schedule interface: production requirements.

The third option treats MPS as schedules of planned-order releases, subject to possible later change. MRP will plan the needed components to produce the end items *if still desirable when the scheduled time comes.* Figure 4-4 shows how this is handled. This views the MPS and MRP as planning tools for components (and other resources), not as part of end-item netting and scheduling; these occur in the execution phase. Under this option, assemblies of end-items *A, B,* and *C* would have to be ordered apart from MRP using final assembly schedules, which are execution tools.

This is the most desirable option. Where MPS items are described by planning BoM and never actually built, this obviously is the correct option. If the end items in the MPS are shippable products or major assemblies of such products, this option will provide flexibility to change final assembly operations to respond to the latest customer product needs.

Locking such execution to MPS planned much earlier blurs the distinction between planning and execution. This leads to "fences" in MPS to limit near-term replanning for differences between customer orders and MPS schedules. The whole approach is a fallacy. This is clear when the implications of fences are considered. Implicitly, two naive (and untrue) statements apply. First, "If we haven't planned it, we can't make it," and second, "If we planned it, we must make it even if we don't need it now." The only true statement is, *"When enough customers' orders arrive to fill the front end of the MPS, stop replanning and execute."*

In make-to-stock manufacturing operations, time-phased order points (TPOP), covered later in this chapter, are often used to control end-item inventory. System designers and programmers are tempted to link planned order schedules from TPOP for end items directly to MRP, in effect making them MPS. This makes data processing efficient but is poor practice, because it links planning and execution tightly. Separation of

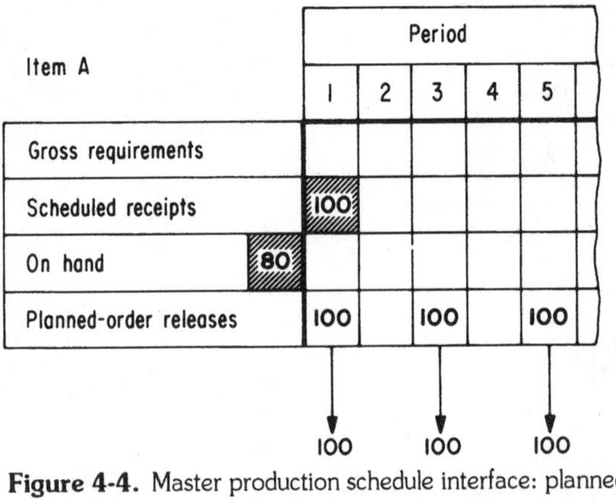

Figure 4-4. Master production schedule interface: planned-order releases.

planning and execution ensures maximum flexibility of operations. A refinement is using firm planned orders to space out TPOP planned orders, leveling the load on the plant and avoiding nervousness in MRP when end-item forecasts or inventories change.

Explosion of Requirements

The power of material requirements planning (and replanning) is the linkage it provides between parent items and components; the latter are needed when the former are to be produced. Parent planned orders establish component gross requirements, to be available at the time the parent order is released for production. The linkage of parent and component inventory records is depicted in Fig. 4-5.

Explosion of requirements in MRP programs from MPS down through component levels follows this logical linkage of inventory records from top to bottom of BoM structure. Gross requirements for items are netted against on-hand and on-order quantities to determine net requirements, which are then covered by planned orders obeying lot sizing and lead time rules. This procedure is carried out repetitively for items on successively lower levels until a purchased component or raw material is reached, where the explosion terminates.

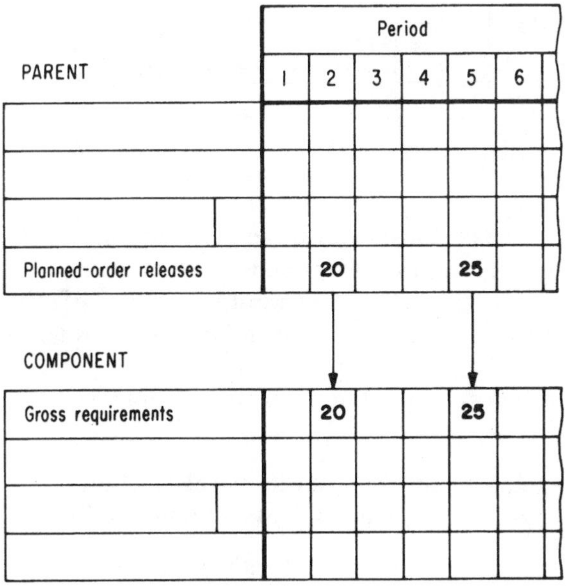

Figure 4-5. Linkage of parent and component records.

Item A - level 1

		Period 1	2	3	4	5	6	7	8	9
Gross requirements		10		15	10	20	5		10	15
Scheduled receipts				14						
On hand	12	2	2	1	-9	-29	-34	-34	-44	-59
Planned-order releases			9	20	5		10	15		

Item B - level 2

		1	2	3	4	5	6	7	8	9
Gross requirements			9	20	5		10	15		
Scheduled receipts										
On hand	28	28	19	-1	-6	-6	-16	-31	-31	-31
Planned-order releases		1	5		10	15				

Item C - level 3

		1	2	3	4	5	6	7	8	9
Gross requirements		1	5		10	15				
Scheduled receipts										
On hand	8	7	2	2	-8	-23	-23	-23	-23	-23
Planned-order releases			8	15						

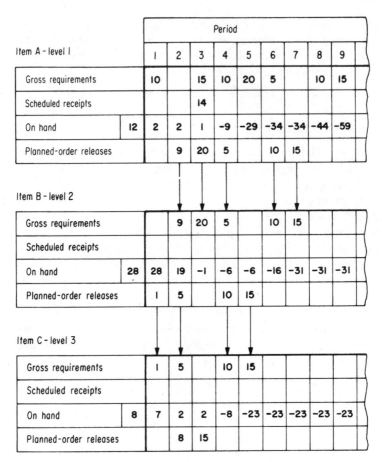

Figure 4-6. Explosion of requirements.

The results of a requirements explosion for three items on consecutive levels are illustrated in Fig. 4-6. In this example, components *B* and *C* do not have multiple parents. If they did, they would require different handling, discussed later in this chapter. The handling is also slightly different in net change MRP programs, explained in Chapter 5.

The computation of component requirements in MRP is complicated by six factors:

1. The structure of the product comprising several levels, including raw materials, component parts, subassemblies, and end items or finished products
2. Lot sizing, ordering items in quantities exceeding net requirements

3. Different individual lead times to procure or produce the items that make up the product
4. Time-phasing end-item requirements stated in the master production schedule across a long planning horizon
5. Multiple requirements for an inventory item used in the manufacture of several other items
6. Multiple requirements for an inventory item used on two or more BoM levels of one end item

Product structure is the principal constraint on the computation of requirements. The number of levels in BoM structures, discussed in Chapter 7, is determined by the sequence of steps in which the product is made from raw material to end item, combining components into parents which then become components of a higher-level parent, and so on to the top level. Product depth (number of levels in its structure) is a major factor in the scope and duration of the MRP data processing job and in the work of storing and maintaining BoM.

Each individual item exists as a uniquely identified physical entity (a unit of raw material, a component part, or a subassembly) and also exists physically built into inventories of one or more parents. In these, it has lost its individual identity. For example, quantities of gear C in Fig. 2-1 in Chapter 2 would be carried on their own inventory record, and *additional quantities* would be part of (already-assembled) gearbox *B* and transmission *A*, without being identified as gear *C*.

MRP assumes that each parent item in inventory is complete with all components. It also assumes that released orders to make more of a parent will be accompanied by the correct quantity of each component. Additional components, therefore, will be needed only when more parents have to be made. Hence, *planned orders for each parent determine gross requirements for its components*. BoM for parents in inventory or work-in-process make it possible to determine the number of components in them, if this is desirable. Product recalls and mandatory design changes may require this.

The general rule of MRP planning logic is

Net requirements for parent items expressed as lot-sized planned orders become gross requirements for components (taking quantity of component per parent into account) in the same time period that parent orders are to be released.

Determining net requirements for a low-level item (the lowest level carries the highest level number, as explained in Chapter 7) must include consideration of the quantity present under its own identity, as well as all quantities of the item now present in all of its parent and higher-level items.

The following example illustrates the basic logic of "netting" requirements.

100 trucks *X* are to be produced, and the following components are on hand (or already on order): Transmission *A:* 2; Gearbox *B:* 15; Gearbox *C:* 7; Forging blank *D:* 46.

What are the net requirements for these items? Without thinking carefully, the answer might be:

Item A:	100 – 2 = 98
Item B:	100 – 15 = 85
Item C:	100 – 7 = 93
Item D:	100 – 46 = 54

This, however, is incorrect, because it ignores quantities of items *B, C,* and *D* already present in their parent items. The correct netting logic is:

Quantity of trucks *X* to be produced:	100
Transmissions required (gross):	100
Transmissions in inventory:	2
Net requirements, transmission *A:*	98
Gearboxes required for 98 transmissions (gross):	98
Gearboxes in inventory:	15
Net requirements, gearbox *B:*	83
Gears required for 83 gearboxes (gross):	83
Gears in inventory:	7
Net requirements, gear *C:*	76
Forgings required for 76 gears (gross):	76
Forgings in inventory:	46
Net requirements, forging *D:*	30

This can be verified as follows:

Quantity of trucks to be produced: 100		
Quantity of forging *D* that will be consumed:		100
Inventory of *D* itself:	46	
Inventory of *C* containing *D:*	7	
Inventory of *B* containing *C* (includes *D*):	15	
Inventory of *A* containing *B* (includes *D*):	2	
Total of all *D*'s available:	70	
Net requirements for more *D*'s:	30	
Totals:	100	100

As computations of net requirements proceed from top to bottom of product structures, level by level, this flushes out (accounts for) component

D "hiding" in higher-level items *A*, *B*, and *C*. If netting proceeded upward through BoM, the where-used traces would lead into other branches that do not apply to product *X*. For example, an additional quantity of item *D* might be found in parent item *Y*; if it entered into the netting process, the final net requirements for item *D* would be understated, because item *Y* is not used in the manufacture of truck *X*.

The downward progression from one product level to another just described is called an *explosion*. Net requirements are developed by applying quantities of each item on hand and on order to meet gross requirements for the item at each level. This netting process may seem laborious but it cannot be circumvented or shortcut; *net requirements on the parent level must be determined before the correct gross and net requirements on the component level can be determined.*

To cover net requirements, the MRP program develops a time-phased schedule of planned orders for each item, including orders, if any, to be released immediately plus orders scheduled for release in specified future periods. Planned order schedules will be revised by the computer when the MRP program is rerun if there have been changes in any of the parameters involved (BoM, MPS, on-hand inventory, etc.). Planned-order quantities are computed using lot-sizing rules (see Chapter 6) specified by the system user for each item.

MRP netting applies existing on-hand quantities to meet item gross requirements. MRP also reevaluates the validity of the timing of any released (open) orders that may now be needed earlier or later than previously scheduled. Good practice leaves to planners the decision about rescheduling open orders; computers simply send a message recommending this. In its entirety, the information on item requirements and coverage that MRP generates is called the *material requirements plan*.

The format and terminology used in figures in this book are those commonly found in MRP programs everywhere. MRP has been lauded for its ability to "make the logic obvious," but the terms used are not obvious and are often misunderstood by new users. Meanings would be clearer using self-explanatory terms such as the following:

Old term	New
Gross requirement	Require
Scheduled receipts	Will get
On-hand or in-stock	Have now
Planned order receipt	Need
Planned order release	Start

MRP works with three types of orders, planned, released (scheduled receipts), and firm planned, and handles them quite differently. The differences are summarized in Fig. 3-9 in Chapter 3.

Planning with Lot Sizes

Order quantities (lot sizes) also are necessary in the requirements computation. This is another reason why the top-to-bottom, level-by-level procedure must be followed. In the preceding example, a tacit assumption was made that parent items *A, B,* and *C* will be ordered in quantities equal to their net requirements. Lot sizing invalidates this assumption, because gross requirements for components are equal to the planned order quantities of their parents.

If gear *C* production order quantities must be multiples of 5 (because of some consideration in the gear-machining process), the net requirement of 76 will have to be covered by a planned order for 80. This will increase the gross requirement for forging blank *D* correspondingly, as illustrated in Fig. 4-7. When the planned order for 80 gears is released, 80 forging blanks will have to be issued.

The specific technique used to determine lot sizes (order quantities) for an item is a major factor in calculating gross requirements for its components. To enable MRP to carry out its explosion, the formulas or rules for lot sizing must be accessible to or part of its computer program. Lot-sizing techniques are presented in detail in Chapter 6.

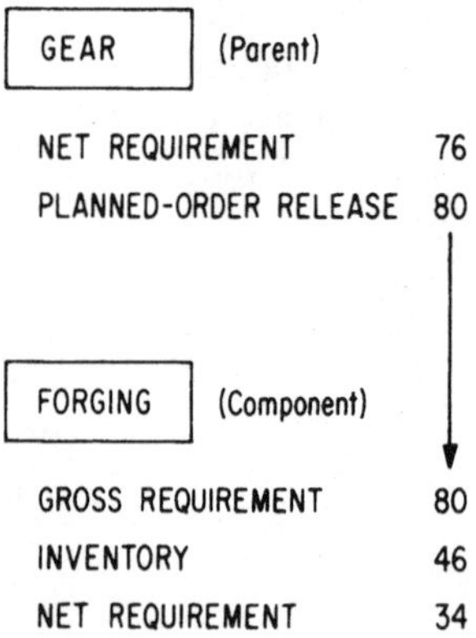

Figure 4-7. Derivation of gross requirements.

Recurring Requirements

Over the MPS planning horizon, there will usually be several requirements in different periods for end items and practically all components. This is another complication in the MRP explosion process. The earlier example of truck *X* ignored requirements for additional lots beyond the 100 specified. If there is a preceding lot of 12 trucks *X* (Lot 1) planned, the net requirements for the two lots, using the lot-sizing rule for gear C of multiples of 5, would be calculated as follows:

	Lot 1	Lot 2
Transmission *A*		
Gross requirements:	12	100
Inventory:	2	0
Net requirements:	10	100
Gearbox *B*		
Gross requirements:	10	100
Inventory:	15	0
Excess available for Lot 2:	5	
Net requirements:	0	95
Gear *C*		
Gross requirements:	0	95
Inventory:	0	7
Net requirements:	0	88
Planned order quantities:	0	90 (Lot-size rule)
Forging blank *D*		
Gross requirements:	0	90
Inventory:	0	46
Net requirements:	0	44

The requirements for Lot 1 and the lot-sizing rule have increased net requirements of all components to make Lot 2. The net requirement for the forging blank, for instance, increased from 34 to 44; this increase of 10 does not match the increase of 12 trucks because of the lot-sizing rule (reorder in multiples of 5) applied to gear *C*.

When determining net requirements, MRP applies component inventories to meet gross requirements *according to the sequence of end-item lots*. If this sequence subsequently changes, as it often will in practice, component inventories must be reapplied and requirements recalculated. A change in end-item lot sequence affects not only the timing but also the quantities of requirements, as can be seen in the forging blank *D*. Had Lot 2 preceded

Lot 1 and the lots represented two different models of trucks that use the same transmission, the net requirement of 30 *D*s for Lot 2 (developed in the first example) would have been correct and that for Lot 1 would be 15, making a total of 45, which is different from the 44 determined for the reverse sequence.

MRP handles sequence changes by time-phasing requirements in the chronological order of the latest MPS. Such changes will constitute a severe problem in MRP *if the MPS is constantly revised for different sequences of end-product production*; this will cause rescheduling of dozens or hundreds of component orders. The computer is capable of coping with such nervousness, but plants and their suppliers are not. Chapter 10 discusses some solutions to this problem.

Handling Common Components

MRP calculations of requirements must include common usage of a component by several parents. This is found frequently in practice, particularly at lower levels of bills of material. The forging blank *D* in the earlier example may be used to make many other gears besides *C*.

In order to determine the net requirements for common-usage items correctly, gross requirements stemming from all parent-item planned orders must first be determined. In addition to dependent demand generated by parent items, a component also may be subject to independent demand if sold as a service part; this independent demand must be added to gross requirements *so that these represent a complete set of demands originating from all sources at different points in time*. This is illustrated in Fig. 4-8.

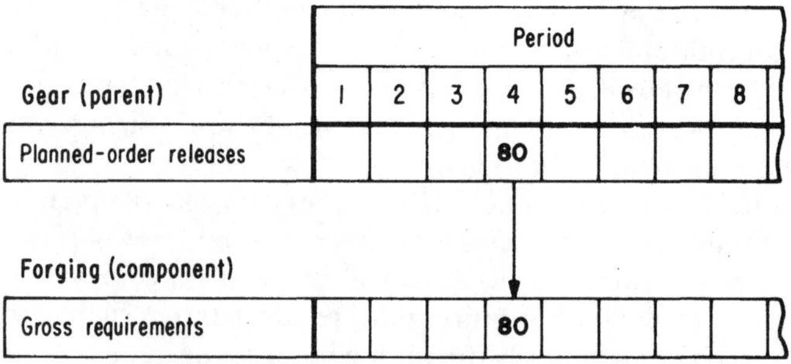

Figure 4-8. Gross requirements originating from multiple sources.

Level-by-level computation of requirements by MRP minimizes the problem of handling multiple-parent demands. If a component appears on only one level in the product structure, all of its parents will be on the next higher level. Because all items on each level are processed before the next level below, all parent planned orders are developed before gross requirements are placed on components. Some components, however, may appear on two or more different levels, adding another complicating factor.

Handling Multilevel Items

Many components are on several BoM levels because their parents appear on different levels. There are two cases: first, an item may appear on different levels in the structures of different end items that use it in common, as is seen for item *A* in Fig. 4-9, or second, it may appear on two or more levels in the structure of one end item, as *A* does in Fig. 4-10. Both conditions may exist at the same time.

A major problem for such items would be having to recalculate requirements every time a parent appears on another level. This would be necessary also for components of such components, requiring multiple retrievals of component records from storage during the requirements explosion, slowing it significantly. This problem is avoided by employing *low-level coding.*

The lowest level at which any item appears in any bill of material is identified by the bill processor program described in Chapter 3, which makes this "low-level code" part of the item's record. In the level-by-level explosion, netting for each item is delayed until the explosion reaches the lowest level on which it appears. At that time all possible gross requirements from parent items at all higher levels have been collected, and netting calculations may proceed. Low-level coding for item *A* is illustrated in Fig. 4-11.

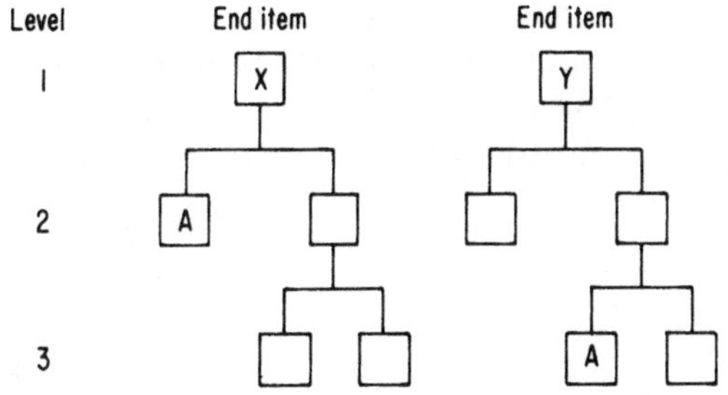

Figure 4-9. Existence of common components on different levels.

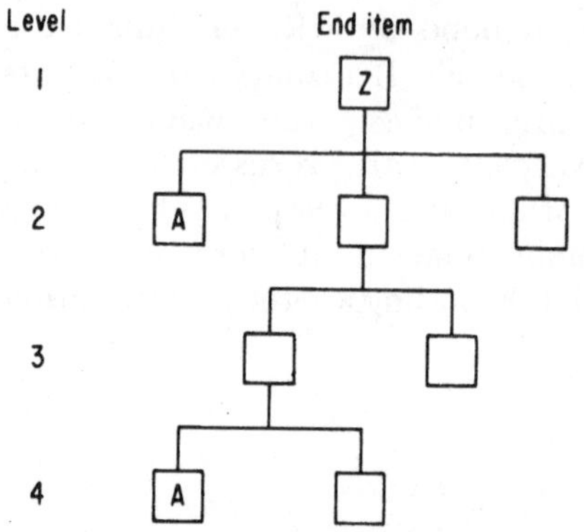

Figure 4-10. An end-item's components on multiple levels.

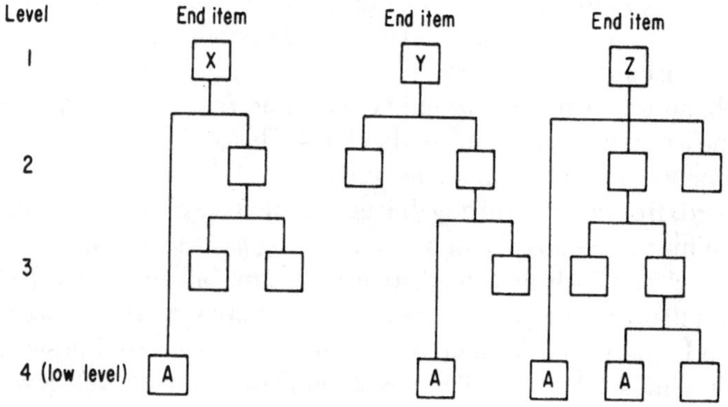

Figure 4-11. Low-level coding.

There is no logical requirement for low-level codes in MRP; their absence will not prevent the program from arriving at correct results. Use of low-level codes yields higher computer-processing efficiency, which usually outweighs the cost of developing and maintaining the coding. Practically every bill-of-material processor software program does this automatically.

Time Phasing

One of the oldest inventions of human civilization is the division of time's continuous flow into increments suitable for measuring its passage, and the construction of calendars to provide a picture of future important events. Our Gregorian calendar serves satisfactorily for most social purposes but has serious deficiencies for inventory planning and production scheduling. It does not employ a decimal base, months have an uneven number of days, the pattern of holidays is irregular, and workdays are not distinguished from others.

To avoid these problems in both manual and computer-based planning systems, special Shop Calendars are used. All serially number days or weeks consecutively; typically, weeks are given two-digit designations (00–99) and working days three-digit designations (days 000–999). The former covers a 100-week and the latter a 1000-day scheduling "year." Such calendars are programmed into computers and can be distributed on cards or placards to people needing them or, better, translated by computers into calendar dates before being shown to people.

Figure 4-12 shows a numbered-week calendar with the "year" divided into four-week "months" designated by period numbers. To convert weeks into periods, divide by 4 and round up; to change periods into weeks, multiply by 4 to get the last week of period.

Figure 4-13 shows a numbered-day calendar, the popular 1000-day, or *M*-day, shop calendar, which counts only working days. Using this, one week normally is equated with five working days. If an item had a five-week manufacturing lead time, 25 days would be subtracted from its due date serial number to arrive at the order release date number. The actual span

PERIOD	WEEK	MON	TUES	WED	THUR	FRI	SAT	SUN	MONTH
24	**93**	1	2	3	4	5	6	7	JAN
	94	8	9	10	11	12	13	14	
	95	15	16	17	18	19	20	21	
	96	22	23	24	25	26	27	28	
25	**97**	29	30	31	1	2	3	4	FEB
	98	5	6	7	8	9	10	11	
	99	12	13	14	15	16	17	18	
	00	19	20	21	22	23	24	25	
1	**01**	26	27	28	1	2	3	4	MAR
	02	5	6	7	8	9	10	11	
	03	12	13	14	15	16	17	18	
	04	19	20	21	22	23	24	25	

Figure 4-12. A numbered-week scheduling calendar.

MONTH	MON	TUES	WED	THUR	FRI	SAT	SUN
JULY	1 / 989	2 / 990	3 / 991	④	5 / 992	6	7
	8 / 993	9 / 994	10 / 995	11 / 996	12 / 997	13	14
	15 / 998	16 / 999	17 / 000	18 / 001	19 / 002	20	21

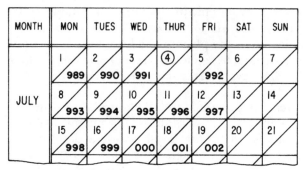

Figure 4-13. M-day scheduling calendar.

of time would be more than five weeks when holidays or plant vacations occurred.

The two types of calendar can be combined in various ways. An *M*-day calendar can also have its (Gregorian) weeks numbered, either without exception or skipping plant vacation weeks. A numbered-week calendar can also identify days by adding a third digit to identify a specific weekday (952 is the second day of week 95). This might mean Tuesday or the second working day, depending on the specific design of the calendar.

Shop calendars make scheduling arithmetic straightforward. They allow points in time to be identified and lead times to be offset by means of simple addition or subtraction. Time-phased data in computer-based planning and control systems are derived from an internal computer shop calendar. Better systems have internal programs to translate from shop to Gregorian calendars and back, avoiding human errors.

MRP inventory status data are time-phased by associating them with days, weeks, or (occasionally) months. The specific method of time phasing employed will determine the way internal arithmetic is carried out by the system, as explained later in this chapter. However, the method of displaying time-phased data (on paper or on a CRT screen) can be selected independently. A given day can be converted to its respective planning period for display and a planning period can be expressed as one of its constituent days, usually its starting day.

There are two alternatives:

1. Using date/quantity data
2. Using time-buckets

In the first, data are displayed in vertical arrays; in the second, horizontal arrays are used. A few users prefer the "bucketless" display in which data are arranged vertically:

Part No.	Requirements	Date
2320	20	June 10
5630	12	May 22
4753	45	July 13

The most common format is the "bucketed" horizontal display used throughout this book. The vertical format lacks its ability to "make the logic obvious" for easy user understanding and action, and also gives an appearance of precision that is spurious. Most MRP programs are designed with requirements and inventory data entered, stored, and internally processed by date/quantity but displayed in time-bucket format.

Time-buckets for periods spanning more than a single day, while very satisfactory for planning requirements, represent a coarse division of time for execution activities. To bridge this gap, a more precise meaning of the timing of data assigned to time-buckets is fixed by user convention. The logic of MRP programs must incorporate whatever time-phasing conventions are specified by the system designer covering:

1. The timing of an event
2. The representation of an activity
3. The size of the bucket

An event is associated with a single point in time. When data representing an event, such as a future receipt of material, are assigned to a time-bucket, it is not self-evident exactly when, during the period, the event is scheduled to occur. One of the following user conventions normally fixes the day of the event in each period for planning purposes:

1. First day
2. Midpoint
3. Last day
4. Any time

Any activity is associated with two points in time—its beginning and end—but one time-bucket can represent only one or the other; what it represents is fixed by convention. An order for a quantity of an inventory item involves two events (Start and Finish) and one activity (Lead Time), illustrated in Fig. 4-14. Time-phasing an order using a single time-bucket allows only one of the two events to be represented; the other will merely be implied or be recorded in some subsidiary record.

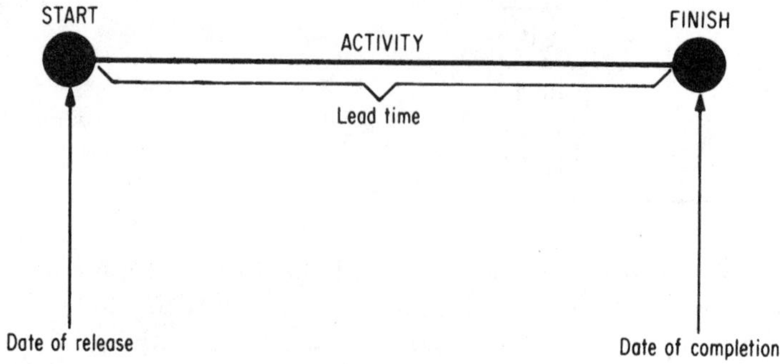

Figure 4-14. The timing of an order.

For an open (previously released) order, its future completion (due) date is more important than its date of release, which is now past. Buckets labeled "Quantities on order," like those in Fig. 4-15, show the timing of scheduled order completions (50 units in period 3, 75 in period 6).

In the case of planned orders scheduled for future release, the timing of the releases is of first importance and the buckets show those. Figure 4-16 shows that 60 units are planned for release in period 2 and another 50 in period 6. To minimize errors in interpreting such data, the wording of the display line is important. The words used in the two figures are ambiguous; more commonly, "Scheduled receipts" and "Planned order releases" are used. As commented earlier, some think "Will get" and "Start" would be even better, but these are not generally accepted.

Bucket size, the length of the period represented by one time-bucket, is also fixed by user convention. A bucket might represent one week, 5 working days, 10 working days, a four-week period, or (rarely) a month. In most MRP programs, time-buckets represent weeks.

Although not recommended, it is possible to assign different time values to buckets in one row. For example, the first 26 buckets might represent weeks, the next 6 buckets, months, and the next 4, quarters, as illustrated in Fig. 4-17. In such a mixed system, passage of time distorts the pattern and must be compensated for either by letting the number of weekly buckets fluctuate between 23 and 26, or by overlapping the last weekly

	Period							
	1	2	3	4	5	6	7	8
Quantities on order			50			75		

Figure 4-15. The timing of order completions.

	Period							
	1	2	3	4	5	6	7	8
Planned-order quantities		60				50		

Figure 4-16. The timing of order releases.

			Period													
	1	2	23	24	25	26	JUL	AUG	SEP	OCT	NOV	DEC	I	II	III	IV
Gross requirements	10	15	5	20	10	10	40	30	50	50	50	30	120	150	150	100

Figure 4-17. A row of buckets with different time values.

buckets with the first monthly bucket. For example, gross requirements for week 27 might be shown separately from those for the month of July, although they would be contained in the total shown for the latter. Computers can handle the needed logic and calculations, but people can be badly confused by such complications. Interest in them fades as planning horizons are shortened.

Inventory Status Timing

MRP evaluates the status of each inventory item to determine planned-order coverage and correct released-order action. The elements of timed inventory status are

1. Quantities on hand
2. Quantities on order
3. Gross requirement quantities
4. Net requirement quantities
5. Planned-order quantities

These status data can be divided into two categories:

1. Inventory data
2. Requirements data

Inventory data include on-hand and on-order (work-in-process) quantities and their timing; these data are reported to the MRP program and can be verified by comparison with item inventory files and open order records. Requirements data are the quantities and timing of gross requirements, net requirements, and planned-order releases; these data are computed by the system and can be verified only through recalculation. The mechanics of updating inventory status data are described in Chapter 3.

Time-Phased Order Points

The logic and programming of time-phased order points (TPOP) are identical with those of MRP. This technique is used for planning and controlling items with independent demand that must be forecast. It is a preferred alternative to order point replenishment techniques for four reasons:

1. It provides information on future planned orders; this provides data for planning all needed resources.
2. It permits replanning of requirements; this keeps relative priorities valid for all shop orders.
3. It links planning for independent and dependent demand for items with both types.
4. It allows planning for known lumps of future demand.

The logic and mechanics of order point inventory planning are covered in Chapter 2. The following example of a service part supplied from factory stock compares OP and TPOP:

Period-demand forecast (F) = 17

Lead time (L) = 2

Safety stock (S) = 100

Order quantity (Q) = 50

On-hand (I) = 170

On-order (O) = 0

Order quantity (OQ) = 50

Order point = Demand during lead time + safety stock

$$OP = (F \times L) + S = (17 \times 2) + 100 = 134$$

A new replenishment order for 50 will be released when future issues drop the quantity on hand plus on order below 134 units.

Figure 4-18 shows TPOP applied to this item. The forecast (17) is projected over the entire planning horizon as period gross requirements. The current quantity on hand (170) is projected to drop below safety stock (100) in period 5, and a replenishment order (50) is planned to arrive then. Offsetting two weeks for lead time, the planned-order release is scheduled in period 3. In period 8, the quantity then on hand (34 + 50 in Lot 1) will again be less than 100, and another planned order, Lot 2, is scheduled for release in period 6 to protect safety stock.

The TPOP results for Lot 1 are identical to those with OP; the order point of 134 is reached sometime in period 3, triggering the replenishment order. Under TPOP, however, *a number of replenishment orders* can be planned; OP just reacts to release one order at a time. If the forecast, on-hand, or on-order inventory, or other parameters are changed, *TPOP can be updated for both planned and released orders.* In addition, planned order release data can be fed into MRP to plan the item's components.

Many service parts are subassemblies or are manufactured from at least one lower-level component. Even if they are no longer in current production, TPOP correctly determines requirements for components of such service parts so that they can be *linked directly to MRP requirements data.*

Actually, the term *time-phased order point* is a misnomer; no order point is calculated or used for any purpose. MRP logic is used to signal ordering at the right time. The MRP program requires no modification to implement time-phased order points. Forecasts are fed to it like MPS, and it handles netting and planned order calculations the same as for any dependent demands on components for which safety stock is specified.

Open orders for items under TPOP control will continually be rescheduled, backward or forward, as forecasts are revised. This will keep their due dates and relative priorities valid, and it can be argued that TPOP works

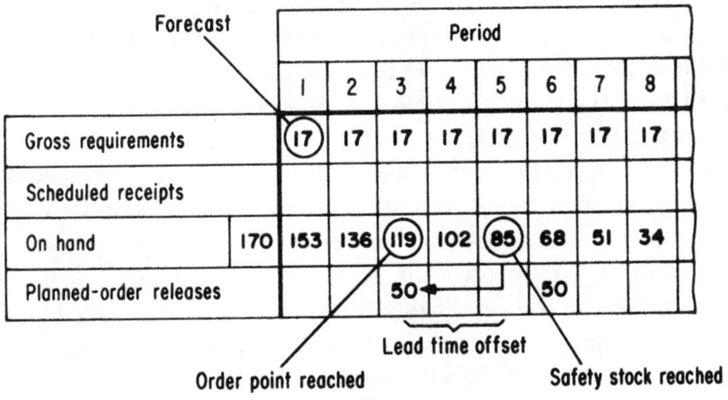

Figure 4-18. Time-phased order point.

well even with poor forecasts. It also has been suggested that close monitoring of forecasts via tracking signals is less important. This ignores the effects of nervousness on good execution by suppliers and plants. The MRP programs cope easily with myriad changes; production cannot.

Time-phased order points also allow a *known future "lump" of demand* to be correctly entered and processed; conventional order point cannot handle this. For example, a service part with forecast usage of 25 units per period is ordered by a customer who wants 200 delivered at a specific future date. Using conventional order point, should an order for 200 be released immediately, even though the lead time is less than the delivery time? Or should it be released at the proper later time and, if so, what special procedure will ensure that the future order release date is met?

OP by itself, of course, will reorder when inventory drops to the order point quantity; the large order may trigger this if entered immediately and, in addition, may increase the normal order quantity. This will increase on-hand plus on-order inventory and will tend to delay the next replenishment order until the 200 units are shipped. By then it may be too late to meet regular service-part demands from other customers.

TPOP eliminates all of these problems. The order for 200 is simply entered as a gross requirement in the appropriate time-bucket. If it is deemed to be an extra to the normal forecast demand, it is added to the forecast for that period. If it is believed to be part of regular demand, period forecasts are adjusted to avoid overplanning replenishment.

Determining Gross Requirements

The term *gross requirements* has a specific meaning in MRP; it is *the quantity of an item needed to support the processing of a parent order or orders.* This may or may not be the total quantity of the item that will appear in the end product. For example, in the example earlier, 100 trucks are to be built, each truck contains one forging blank *D,* and the gross requirement for *D* might be thought to be 100. For activities other than MRP (cost accounting and capacity planning), this is true.

In MRP, however, the question is not what quantity of a component will be shipped with the product, but what *minimum* quantity will have to be procured or manufactured to build each lot of product in the MPS or parent('s) schedule. In the example, the gross requirement for forging blank *D* was 76 and the net requirement 30. The gross requirement for *D* would have been 100, however, had there been no inventories of gears, gearboxes, or transmissions. In MRP the gross requirement is the *demand at each item level, rather than demand at product or master production schedule end-item level.*

Determining Net Requirements

As stated earlier, an item may have dependent demand from several parent items and may also have independent demand from sources external to the plant. Item *A*, previously shown in Fig. 4-8, has both types of demand. Its gross requirements are reproduced in Fig. 4-19, which shows that there are 23 units of this item on hand and 30 more on order, due in period 3. The common term "Scheduled receipts" is used in the figure, indicating a released, not a planned, order.

The logic of the net requirements computation is

Gross reqts. – On-hand – Sched. recpts. = Net reqts.

A negative net requirement, when the sum of quantities on hand and on order exceeds gross requirements, is considered to be zero. As shown in Table 4-1, data in MRP are time-phased and the netting calculation is performed successively for each period; unapplied inventory is carried forward into the next period(s).

	Period 1	2	3	4	5	6	7	8	Total
Gross requirements		20		25		15	12		72
Scheduled receipts			30						30
On hand: 23									

Figure 4-19. Time-phased status before net requirements computation.

Table 4-1. Calculation of Net Requirements

Period	Gross requirements	Scheduled receipts	On hand		Result	Net requirements
1	0	–0	–23	=	–23	0
2	20	–0	–23	=	–3	0
3	0	–30	–3	=	–33	0
4	25	–0	–33	=	–8	0
5	0	–0	–8	=	–8	0
6	15	–0	–8	=	7	7
7	12	–0	–0	=	12	12
8	0	–0	–0	=	0	0
Totals	72	30				19

The correctness of the computation can be verified easily:

Total gross requirements:		72
On hand:	−23	
Total scheduled receipts:	−30	
Total available:		−53
Total net requirements:		19

Figure 4-20 shows these data in time-phased record format. Period net requirements are equal to gross requirements, except in periods during which inventory runs out or is increased by a new receipt of the item.

An alternative method of calculating net requirements is to project the quantity on hand into the future, period by period; the first negative value then represents the first net requirement. Differences between successive negative values equal the net requirements in the respective later periods. The logic of the on-hand/net-requirements computation is

Balance on hand at the end of a given period

plus **quantity on order due in the next period**

minus **gross requirements of the next period**

equals **balance on hand at the end of the next period.**

In this method, "negative on-hand" is understood to equal net requirements recorded cumulatively, as shown in Table 4-2. This method provides the simpler display shown in Fig. 4-21; it combines the projected on-hand and net-requirements data in a single row of time-buckets. Negative on-hand data indicate net requirements; the first is 7 in period 6. The cumulative negative on-hand, (19) in period 8, equals total net requirements over the horizon. Negative on-hand data show what will happen to the item's inventory if another replenishment order is not

		Period								Total
		1	2	3	4	5	6	7	8	
Gross requirements			20		25		15	12		72
Scheduled receipts				30						30
On hand	23									
Net requirements							7	12		19

Figure 4-20. Time-phased status after net requirements computation.

Table 4-2. Alternative Calculation of Net Requirements

Period	On hand at beginning of period	Scheduled receipts	Gross requirements		On hand at end of period
1	23	+0	–0	=	23
1	23	+0	–20	=	3
3	3	+30	–0	=	33
4	33	+0	–25	=	8
5	8	+0	–0	=	8
6	8	+0	–15	=	–7
7	–7	+0	–12	=	–19
8	–19	+0	–0	=	–19

planned. Although simpler, this display does not make the logic obvious, and users must become familiar with their real meanings to use the data properly.

Safety Stock Effects on Net Requirements

MRP can include planning of safety stock, but this is not recommended and is not common practice. When included, the quantity of safety stock is either subtracted from the on-hand quantity or added to gross requirements; the former is common. Either alternative produces the same result, increasing net requirements and often shifting the first net requirement one or more periods forward. Including 2 units of safety stock, the projected on-hand and net requirements quantities in Fig. 4-21 would change, as shown in Fig. 4-22.

		Period								Total
		1	2	3	4	5	6	7	8	
Gross requirements			20		25		15	12		72
Scheduled receipts				30						30
On hand	23	23	3	33	8	8	–7	–19	–19	–19

Figure 4-21. Alternative method of net requirements display.

Safety stock: 2		Period								Total
		1	2	3	4	5	6	7	8	
Gross requirements			20		25		15	12		72
Scheduled receipts				30						30
On hand	23	21	1	31	6	6	-9	-21	-21	-21
Net requirements							9	12		21

Figure 4-22. Net requirements after safety-stock deduction.

When safety stock is planned at the item level, MRP logic attempts to "protect" it from being used. To the extent that this succeeds, it creates "dead" inventory. Other harmful effects result when people know this cushion exists and relax efforts to get orders completed on time or lose confidence in MRP when they see that order due dates are not valid. Chapter 11 contains more discussion of safety stock in MRP.

Coverage of Net Requirements

Net requirements for an item indicate impending shortages; the item will not be available at that time unless some action is taken. Displays like the one in Fig. 4-21 show clearly when more material is needed. Assuming an adequate planning horizon and effective execution, MRP detects such shortages sufficiently in advance to allow coverage to be planned to prevent them.

Open (released) orders are deducted from gross requirements in calculating net requirements. Coverage of net requirements involves setting planned orders in the proper time periods (quantities are set by lot-sizing rules) when they are needed. Generating time-phased planned orders for components is one of the most significant abilities of MRP.

Planned Orders

The principal value of planned orders for an item is providing advance data for a rigorous determination of its component requirements. They also provide "visibility" into the future and data for projections of needed resources including on-hand inventory, purchase commitments, and production capacity.

The timing of component gross requirements is linked directly to time-phased parent planned-order due dates. Planned-order quantities are determined by ordering policies selecting lot-sizing techniques specified by users. Different techniques can be applied to different items or item classes. These are presented in Chapter 6.

MRP covers net requirements with planned orders for future release. Planning horizons should extend far enough into the future so that all components of MPS end items have at least one net requirement and one planned order. Chapter 3 gives details of lead time/planning horizon relationships. To generate a planned order correctly, MRP must determine

1. The timing of order completion (due) date
2. The timing of order release (start) date
3. The order quantity

The *timing of order completion* must coincide with the timing of the net requirements being covered if shortages are to be prevented. Because MRP uses time-buckets, *dates for order completion* will be stated in similar time periods. Two alternatives are possible if more precise timing is desired. First, adopt a convention assigning a specific time (first day or midpoint, for example) to each time-bucket; this is the common way. Second, adopt time-buckets of the same period as desired for order dates (daily, for example); the drawbacks of this approach are presented in Chapter 11.

In the previous example in Fig. 4-22, the first net requirement is 7 in period 6. First-time users often question whether the planned order to cover this net requirement should be scheduled for completion in period 5 or in period 6. This question involves the time-bucket timing convention (discussed earlier), usually either the beginning or the midpoint of the period.

In the first case, the beginning of one period is the same as the end of the preceding period, and order completion would be planned for (no later than the end of) period 5. In the second case, order completion would be planned for period 6, meaning no later than the middle of this period. The time of order completion, of course, determines the time of order release; this will be planned earlier depending on the lead time required and the timing convention used. The *planned order quantity* will be set by the lot-sizing rule selected by users for each item.

Lead Time Offsetting

MRP determines the recommended planned order release timing by offsetting an item's lead time, subtracting the proper shop calendar units (discussed earlier in this chapter) from the date of order completion. For example,

Order completion	week 6
Lead time	– 4 weeks
Order release	week 2

or, alternatively,

Order completion	day 328
Lead time	– 20 days
Order release	day 308

Lead times usually are stated in the same units as the planning period; when different units are used, some minor distortions may occur. If the lead time is eight days, for example, MRP using weekly time-buckets will treat it as two weeks. Lead times of four days or less will be treated as one week. Sometimes this distortion becomes extreme; for example, if subassembly *A*, made from *B*, which is made from *C*, each has a one-day lead time, MRP using weekly time-buckets will offset three weeks, starting *B* and *C* much earlier than needed. To prevent such inflation of the cumulative lead time, the component lead times may be specified as zero. The system will then schedule all three order start and completion dates in the same week.

Figure 4-23 adds a line for "Planned-order releases" to the display in Fig. 4-21 and shows a planned order (arbitrarily lot-sized to cover both net requirements) offset for a lead time of four periods. Part *A* shows the midpoint convention, with week 2 as order release time; Part *B* uses the start-of-period convention and indicates week 1. When the planned order is released, it will become an open order and will appear on the "Scheduled receipts" row. The timing convention of planned-order releases will also determine the timing of scheduled receipts, either week 5 or 6.

Figure 4-24 shows how the display would appear following release of the planned order in period 2. *In this and in all subsequent examples and discussions,*

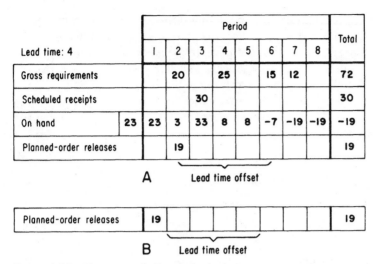

Lead time: 4		Period 1	2	3	4	5	6	7	8	Total
Gross requirements			20		25		15	12		72
Scheduled receipts				30						30
On hand	23	23	3	33	8	8	-7	-19	-19	-19
Planned-order releases			19							19

A Lead time offset

Planned-order releases	19							19

B Lead time offset

Figure 4-23. Alternatives of offsetting for lead time.

		Period 2	3	4	5	6	7	8	9	Total
Gross requirements		20		25		15	12			72
Scheduled receipts			30			19				49
On hand	23	3	33	8	8	12				0
Planned-order releases										0

Figure 4-24. Status following release of planned order.

the midpoint convention will be used, being the more common. Individual item lead times used in MRP, such as the four weeks in the present example, are *planning average estimates* needed to make its calculations possible. Actual lead times of specific item orders will vary widely from order to order around this average. Reasons for this, and ways to minimize such variations, are presented in Chapter 11.

External Component Demands

In the typical manufacturing environment in which MRP would be used, most component demands stem from MPS and are generated internally by the explosion process. Often, some demand for components comes from sources external to the plant (service parts or interplant) or from nonproduction sources within the plant (experimental, design engineering, quality control, or maintenance). The latter demands are usually minimal and sporadic, rarely warranting separate planning. Service parts and interplant demands, on the other hand, often are both significant and recurring. These demands are conveyed to MRP in one or more of the following ways:

1. Orders placed by a service parts warehouse
2. Orders placed by another plant or function
3. Forecasts of service-part demand
4. Planned-order schedules of a service warehouse time-phased order point system
5. Planned-order schedules of another plant's MRP system

Service-part and/or interplant orders are entered into a plant's MRP program by transactions that increase each item's gross requirements in the time-bucket corresponding to the order due date, usually specified by the customer. These are then added to gross requirements stemming from parent planned orders, and the requirements planning process then proceeds normally.

When the originator of service parts orders uses an MRP program, whether it is part of the company supplying the parts or independent of it, the most effective way of transmitting service-part demand is to use time-phased order points. As shown earlier in this chapter, these can be run on MRP programs and will permit the integration of service-part demands into the supplier plant's MRP directly. By furnishing schedules of planned orders developed by TPOP, customers, whether internal or external, can eliminate the need for individual part orders and provide suppliers with more useful information about future requirements.

This method of feeding the output of a customer's TPOP program directly into its supplier's MRP is being applied widely to replenishment of central and regional warehouses and agents' and distributors' inventories. In this process, two time-phased inventory records of the same item are linked in a pseudo parent-component relationship. Figure 4-25 shows how this is done. Service part *X*, in its customer time-phased order point record, acts as a parent item generating gross requirements in the supplying plant's MRP system. Independent demand for the item at the warehouse level is translated, by the processing logic, into independent demand at the plant level.

Item X
Safety stock: 20
Lead time: 1

"Customer" system

Period		1	2	3	4	5	6
Gross requirements		10	10	10	10	10	10
Scheduled receipts							
On hand	48	38	28	18	8	-2	-12
Planned-order releases			30			30	

Item X
Lead time: 4

Supplying plant system

Period		1	2	3	4	5	6
Gross requirements			30			30	
Scheduled receipts			30				
On hand	5	5	5	5	5	-25	-25
Planned-order releases		25					

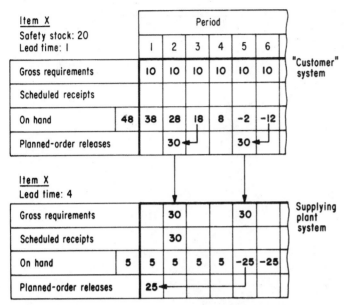

Figure 4-25. Pseudo parent/component relationship.

Because the item in question has two time-phased inventory records, it must have two lead times: one for the parent, equal to the transportation or delivery time between warehouses, and another for the component, equal to the manufacturing time. In Fig. 4-25 these are one and four weeks, respectively.

The discussion in this chapter of MRP's processing logic has focused on characteristics commonly found in practically all such systems. There are two alternatives in operating and updating MRP programs, however, requiring very different procedures. These are called *regeneration* and *net change* and are adapted to different frequencies of replanning. Their description and comparison are the subjects of Chapter 5.

5
Keeping MRP Up-to-Date

Every revision, however necessary, will be carried by weak minds to an excess that will itself need revising.

SAMUEL TAYLOR COLERIDGE

Regeneration and Net Change

There are two alternative ways of keeping MRP up-to-date:

1. Schedule regeneration
2. Net change

The differences between regeneration and net change are the frequency of replanning and what initiates it. *Regeneration is run periodically when all master production schedules, revised or not, are fed to MRP. Net change is run at more frequent, random intervals when inventory transactions are processed.* Regeneration requires long computer runs, albeit at high data processing efficiency; this usually limits the frequency of replanning to weekly or longer cycles. Net change is designed for almost continuous replanning at the expense of data processing efficiency. The principal outputs of these two updating methods are identical in content and the inputs are nearly identical, the exception being a difference in item inventory status maintenance.

An MRP program is either regenerative or net change. There are no other basic alternatives, but a whole spectrum of minor variations are found in practice to meet particular needs of users. Regenerative programs "borrow" some net change features and conversely, net change programs sometimes are used periodically, like regeneration. Net change programs also are regenerated occasionally to purge an accumulation of small errors. To avoid confusion in this chapter, "pure," or "classic" versions of schedule regeneration and net change will be described.

What sets MRP updating programs in motion depends on how they are replanned. Regenerative MRP programs employ batch-processing techniques and replan periodically, typically in one-week intervals; passage of time triggers the process. Net change MRP programs, however, use inventory events (transactions) to initiate replanning, more or less continuously.

Four changes—in requirements, coverage of these, product structure (engineering changes), and planning factors—affect inventory status and must therefore be reflected in replanning. Regenerative MRP programs take a "snapshot" of these factors at the time of each periodic replanning; the implicit assumption is that all relevant changes have been incorporated during the preceding interval; these programs deal periodically with static situations. Net change MRP programs, on the other hand, attempt to deal with dynamic, fluid situations; this requires that changes in any of these four factors be reported to MRP as they occur or at frequent, short intervals.

Regeneration

From its beginnings in the 1960s, conventional MRP used regeneration for replanning. In this,

1. Every end-item requirement stated in the MPS is exploded through all levels in MRP.
2. Every active bill of material is utilized.
3. The inventory and order status of every active inventory item is recomputed.
4. Voluminous output is generated.

Regeneration explodes all requirements in a batch-processing run at a predetermined frequency, using the latest data in MPS and other files. During this run, gross and net requirements for each inventory item are recomputed, each released order status reviewed, and all planned order schedules recalculated. The run is made level-by-level through bills of material, starting at end-item MPS levels and progressing down BoM to purchased materials.

Regeneration involves periodic, batch, sequential data processing and is inherently a massive data-handling task. Changes, if any, in master production schedules, product structures, item inventory and order status, and in planning factors can be accumulated for processing between regeneration runs. Weekly or biweekly replanning cycles are typical.

The operation of regenerative MRP programs occurs in two distinct, alternating phases:

1. Requirements planning (explosion) run
2. Intracycle file updating

The latter consists of frequent reporting of inventory transactions to the program and posting them to individual inventory records. This keeps these records (reasonably) up-to-date for planner inquiries and also readies them for the next replanning run. File maintenance for interim changes in product structure and planning factors, such as lead times and scrap allowances, must be completed also before regeneration.

As mentioned in Chapter 4, two types of data constitute the status of inventory items in MRP, *inventory data* (on-hand and on-order) and *requirements data* (gross and net requirements and planned-order releases). In regenerative replanning, inventory data are maintained by file-update processing of transactions at a relatively high frequency, usually daily. Requirements data are reconstructed or reestablished at a lower frequency set by the regeneration cycle. Outputs of the replanning run that reestablish requirements status are typically printed as reports rather than stored in computer files; if users desire faster inquiry, they can be held in the computer for displays.

In file updating, a transaction adjusts only one item record directly. Some transactions (scrap, for example) may alter the inventory status of a parent and also the requirements status of its components, because of the MRP link between parent planned orders and component gross requirements. In regenerative MRP this poses no problem, since file updating and requirements recalculation are separate activities; inventory transactions never trigger explosions into lower product levels, resulting in a gradual deterioration in the validity of requirements data between successive replanning.

Recapitulating the two points just made:

1. A transaction may change parent item status, but file updating will not cause changes in component status.
2. Requirements data will gradually deteriorate if component status is not modified for parents' transaction changes.

Regeneration Frequency

The frequency of replanning is a critical variable in the use of an MRP program. Regeneration's massive data handling during the requirements planning run dictates that it be done only periodically, at some reasonable intervals. This delay causes MRP to be increasingly out of date between replanning runs. How serious this may be depends on

1. The environment in which MRP must operate
2. The uses made of its output data

Dynamic, volatile environments present serious obstacles to using MRP effectively. In the past, most companies existed in a continual state of change. Customer demands resulted in frequent changes in master production schedules, released order schedules changed sometimes day by day, interplant and field warehouse orders arrived erratically with rush service-part orders, scrap was unpredictable, and a constant stream of mandatory, immediate engineering changes had to be implemented. In such an environment, while there is a strong need for more timely response to change, regenerative MRP programs can replan only periodically; at best, once a week.

In a more stable and predictable environment, regeneration at weekly intervals may function satisfactorily. However, MRP's capability to keep released order due dates up-to-date and valid is of vital importance and may be jeopardized by infrequent replanning. These due dates provide the bases for establishing relative priorities of shop orders for correct sequencing; valid shop priorities require up-to-date order due dates. Replanning MRP on weekly or longer cycles may result in invalid shop priorities. Planning valid priorities is covered in Chapter 9; controlling them is discussed in detail in Chapter 10.

Requirements Alteration

A special feature of some regenerative programs, called *requirements alteration*, is sometimes confused with net change. Requirements alteration is designed to process changes in the MPS between the periodic MRP regeneration runs to get more frequent partial replanning without the effort and expense of a full explosion. Inputs to the program are new period MPS values (not net differences) for end items having changes. The explosion is limited to the changed end items and their components.

Requirements alteration handles only changes in the MPS; it does not react to changes caused by transactions on lower-level items as net change does. It updates the status of only those lower-level components related to MPS end items being processed. As mentioned earlier, the status of all other items deteriorates between regeneration runs. Few MRP programs use requirements alteration; of those that do, some process only changes between regular periodic regeneration runs, which then also process the unchanged portion—often the bulk—of the MPS. A comparison of the actions of regeneration, requirements alteration, and net change is shown in Fig. 5-1.

Net Change Replanning

To make more frequent replanning possible, the delay inherent in regeneration's massive batch-processing, even using today's powerful computers, can be avoided. Net change replanning provides the solution; it has a much smaller scope, shorter duration, and far less output, but uses more total computer time.

Replanning requirements in MRP involves level-by-level explosions. These cannot be eliminated, but net change makes them in high-frequency, consecutive, partial explosions. This minimizes the scope of replanning at any one time and thus permits frequent processing. Making only partial explosions automatically limits the volume of the resulting output.

Net change explosions are partial in two senses; first, only part of the master production schedule is exploded at any one time; and second, the effects of transaction-related explosions are limited to lower-level components.

	Regeneration	Requirements Alteration	Net Change
Inventory, Open Order Files Updated	Periodically	Periodically	Continuously (Short Intervals)
All Master Production Sched. Exploded	Yes	No	No
Frequency of MRP Run	Periodically (Longer Intervals)	Periodically (Short Intervals)	Periodically (Short Intervals)
Effects of Every Transaction Analyzed	No	No	Yes

Figure 5-1. MRP updating actions.

Net change MRP programs can be used in two ways:

1. High-frequency replanning, typically a daily batch
2. Continuous or on-line replanning

Typical practice with net change systems is daily batch transaction processing, requirements and order replanning, and continuous on-line inquiry into the inventory item file. Transactions are accumulated throughout the day and sorted prior to the updating run. For data processing efficiency, other sequential processing techniques may be used, including low-level coding (see Chapters 3 and 4). Transaction processing updates item inventory records and carries out the partial explosions required to maintain interlevel equilibrium as defined later in this chapter.

Net change programs may be used for either daily batch or on-line transaction processing, at the user's option. On-line transaction entry requires data terminals and program software external to the net change program to feed data to it and access it for status data. With this facility, net change keeps data up-to-date as of the last transaction processed. The less delay there is in entering transactions, the more up-to-date it is.

Master Production Schedule Continuity

Net change views the MPS as one basic, continuous plan rather than as separate, successive versions or new plans. The MPS can be updated at any time in net change by adding or subtracting net differences from an end item's previous status. New item schedules are viewed also as net differences in the MPS.

The concept is illustrated in Fig. 5-2. The schedule is envisioned as a scroll unwinding from right to left with passage of time; each bucket in the grid contains either zero or some positive value. MPS extend indefinitely into the future; all buckets beyond the planning horizon having unspeci-

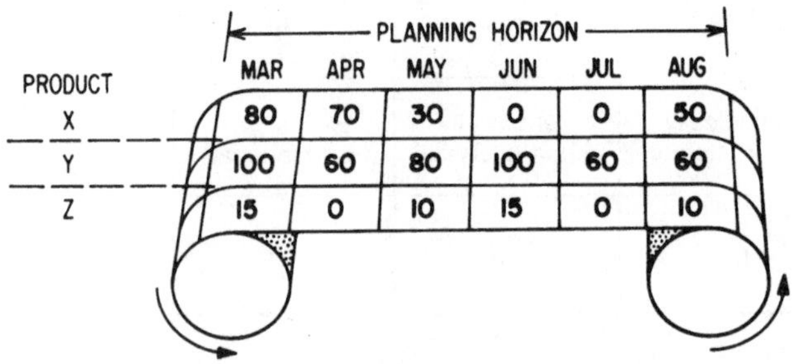

Figure 5-2. Master production schedule continuity.

A

PRODUCT	MAR	APR	MAY	JUN	JUL	AUG	SEP
X	80	70	30	0	0	50	0
Y	100	60	80	100	60	60	0
Z	15	0	10	15	0	10	0

B

PRODUCT		APR	MAY	JUN	JUL	AUG	SEP
X		70	30	0	0	35	40
Y		60	80	100	60	60	0
Z		0	10	15	0	10	15

C

PRODUCT		APR	MAY	JUN	JUL	AUG	SEP
X						−15	+40
Y							
Z							+15

Figure 5-3. Net change between consecutive master production schedules.

fied or unknown contents. Passage of time brings future buckets within the planning horizon, at which time their contents will be revealed as zero or a positive value.

Updating and changing MPS are equivalent under net change; both are done by addition or subtraction of net differences to or from a bucket's previous data. This was pioneered by the American Bosch Company of Springfield, Massachusetts, in a biweekly batch MRP program implemented in 1959; Figure 5-3 illustrates their approach. A six-month schedule made in late February (Part A) and revised in March (Part B) results in the net differences (Part C). These net changes would be processed (exploded) by MRP on the date the new schedule becomes official.

In this example, 15 of 18 MPS buckets within the planning horizon are unchanged, including the complete schedule for Product *Y*. Data processing with net change requires only a fraction of the work regeneration programs would have to perform if all 18 buckets of MPS data, all inventory and open order records, and all bills of material for products *X*, *Y*, and *Z* would have to be accessed and processed.

If the need to reduce the August quantity of Product *X* had been recognized sometime in March, net change could have processed it at that time without waiting for the April schedule. In that case, the net impact on Product *X* of the April schedule would be limited to the addition of 40 units in September.

Updating Item Status

The principle of net change—processing only differences from previous MPS data—extends also to item-status updating. Inventory status in its broader sense can be maintained up-to-date for all items covered by MRP without regenerating all data. Gross and net requirements data need not be recalculated completely, merely modified. In net change, these data are updated in the process of posting inventory transactions to item records. File updating, limited under regeneration to on-hand and on-order data, is expanded to cover all item-status data.

This requires that time-phased requirements data be logically integrated with the traditional on-hand and on-order inventory status data and that a permanent requirements file be created for each item. Unlike regeneration, where these data are only printed, this file is stored in the computer and maintained up-to-date by the net change program. Gross requirements for an item derive from the quantities and timing of parent-item planned order releases, so these data also are stored, either separately or as part of the requirements file.

Two inventory status concepts characterize net change:

1. Record balancing
2. Interlevel equilibrium

Record balance exists when the *projected on-hand quantities correspond to existing gross requirements plus scheduled receipts, and when planned orders have both correct quantity and timing.* The next transaction—receipt, issue, or adjustment—will change the status and may disturb this balance. Net change rebalances the record by recomputing projected on-hand and net requirement quantities and realigning or revising the planned-order releases. All inventory records on file are maintained in balance at all times.

In Fig. 5-4, if scrapping 3 pieces reduced the open order from 23 to 20, net requirements are increased and should move forward in time. To restore balance, the planned-order release will have to be advanced, as shown in Fig. 5-5.

For interlevel (or file) equilibrium, *gross requirements for each component must correspond at all times to the quantities and timing of planned-order releases of its parents.* Moving the planned order for 25 pieces from period 4 to period 2 advances gross requirements of its components 2 periods. To restore equilibrium, the net change program debits period 4 and credits period 2, as shown in Fig. 5-6.

Restoring interlevel equilibrium requires a partial explosion of requirements for items affected. Product structure records identify all components of the parent item; their item records are retrieved and processed to

Lead time: 3
Order quantity: 25

		Period							
		1	2	3	4	5	6	7	8
Gross requirements		10	2		10	13		20	4
Scheduled receipts				23					
On hand	14	4	2	25	15	2	2	-18	-22
Planned-order releases					25				

Figure 5-4. Status before scrap.

Lead time: 3
Order quantity: 25

		Period							
		1	2	3	4	5	6	7	8
Gross requirements		10	2		10	13		20	4
Scheduled receipts				20					
On hand	14	4	2	22	12	-1	-1	-21	-25
Planned-order releases			25 ←	←	←				

Figure 5-5. Status rebalanced following scrap.

Component items	Period							
	1	2	3	4	5	6	7	8
CHANGE IN GROSS REQUIREMENTS		+25		-25				

Figure 5-6. Net change in component gross requirements.

reestablish balance in the status of the individual items. If this changes the planned-order release schedules of the component items, and if these also have lower-level components, the explosion and rebalancing continues down the product structure through all lower levels affected.

This is illustrated in Fig. 5-7. The item-record files in Part A (left portion) of the figure are in equilibrium. Then a customer returned 4 units of assembly *X*, increasing the on-hand quantity from 19 to 23 as shown in Part B; this upset the equilibrium between records *X* and *Y*. Rebalancing *X-Y* upset the *Y-Z* equilibrium, which was restored as shown. In this example, a single transaction triggered an explosion into at least three lower levels. The processing logic illustrated in this example is the same as that used in

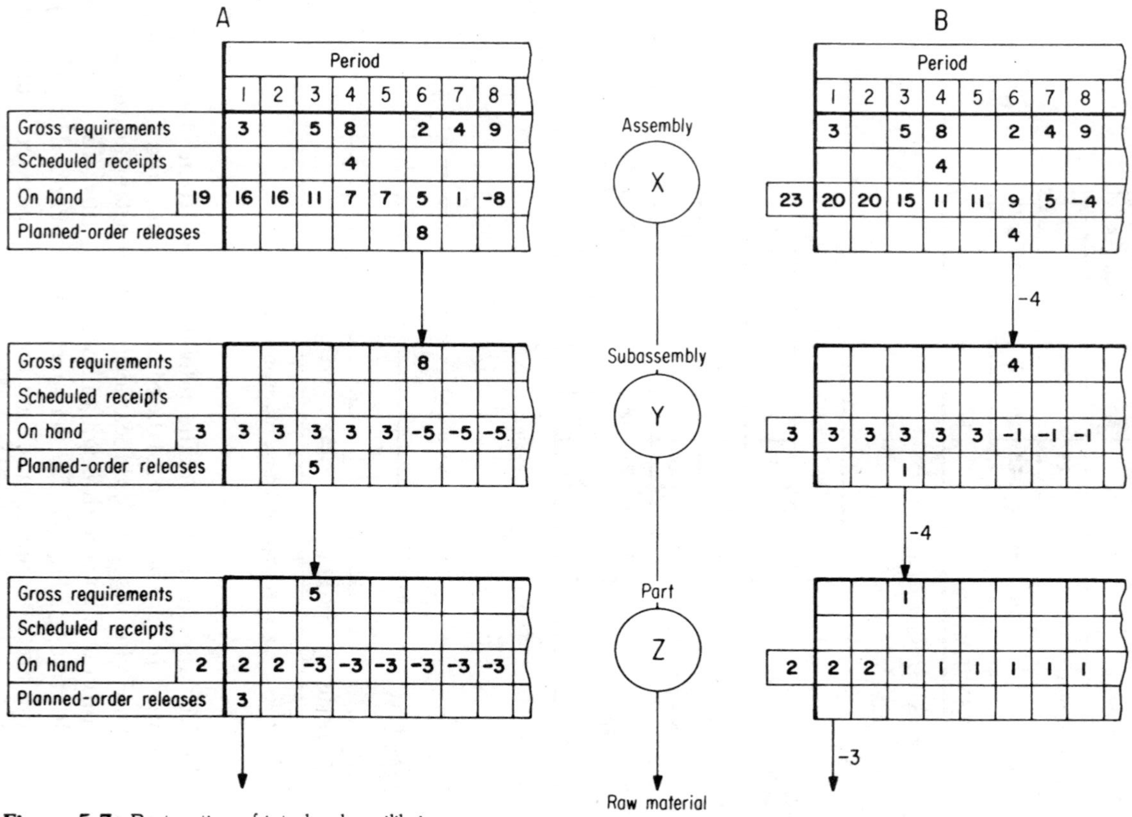

A — Assembly X

		1	2	3	4	5	6	7	8
Gross requirements		3		5	8		2	4	9
Scheduled receipts					4				
On hand	19	16	16	11	7	7	5	1	-8
Planned-order releases							8		

A — Subassembly Y

		1	2	3	4	5	6	7	8
Gross requirements							8		
Scheduled receipts									
On hand	3	3	3	3	3	3	-5	-5	-5
Planned-order releases				5					

A — Part Z

		1	2	3	4	5	6	7	8
Gross requirements				5					
Scheduled receipts									
On hand	2	2	2	-3	-3	-3	-3	-3	-3
Planned-order releases		3							

B — Assembly X

		1	2	3	4	5	6	7	8
Gross requirements		3		5	8		2	4	9
Scheduled receipts					4				
On hand	23	20	20	15	11	11	9	5	-4
Planned-order releases							4		

B — Subassembly Y

		1	2	3	4	5	6	7	8
Gross requirements							4		
Scheduled receipts									
On hand	3	3	3	3	3	3	-1	-1	-1
Planned-order releases				1					

B — Part Z

		1	2	3	4	5	6	7	8
Gross requirements				1					
Scheduled receipts									
On hand	2	2	2	1	1	1	1	1	1
Planned-order releases									

Figure 5-7. Restoration of interlevel equilibrium.

regeneration, but transaction-triggered explosions are a unique characteristic of net change.

The Role of Transactions

Getting maximum use of MRP requires maintaining interlevel equilibrium; the effects of inventory transactions must be fully processed through all lower bill-of-material levels as part of updating item records for these transactions. This modifies the level-by-level processing described in Chapter 4. Based on discussion in that chapter, compare Fig. 5-8 with Fig. 5-9, which illustrates the explosion path in net change.

Net change makes no distinction between file updating and requirements replanning; inventory accounting and requirements planning are fused into a single function. For each transaction, net change examines all affected items and brings their records into equilibrium. Therefore, *transactions may be entered in random sequence and at random times.* Such

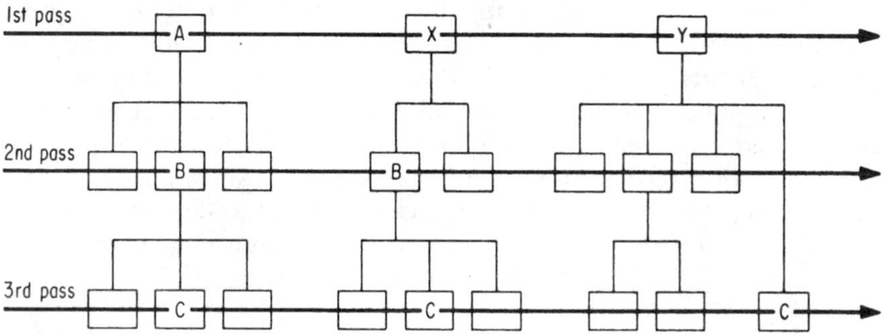

Figure 5-8. Level-by-level processing.

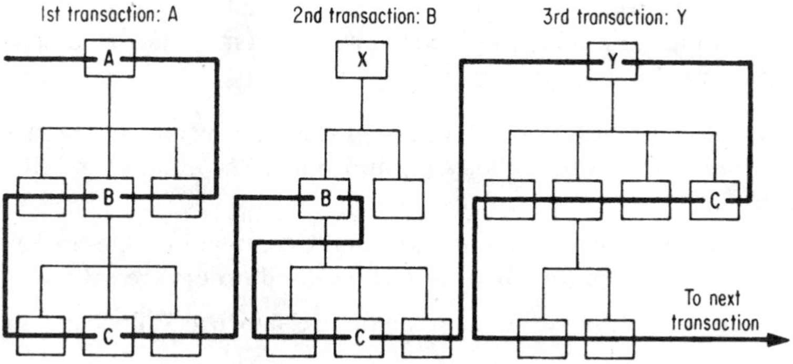

Figure 5-9. Modified level-by-level processing.

random handling, of course, requires additional data processing, and thus causes inefficiencies. Compared to the path in Fig. 5-9, regeneration processing in batches would proceed level-by-level, handling *A* and *Y*, then *B*, and then *C* in turn.

Net change, being transaction-driven, can handle transactions entered through remote on-line terminals. It updates both inventory status and requirements data; any entry (input) posted to an item record that affects time-phased status data acts as a transaction. The following inputs are viewed and treated as inventory transactions by net change programs:

1. Data in master production schedules
2. Gross requirements revisions caused by changes in parent planned-order release schedules generated internally
3. Gross requirements additions of external, direct entries to lower-level item records (e.g., a service-part order)
4. Traditional inventory transactions

Some classes of changes normally are not handled in net change programs. Design changes altering product structures and changes in planning factors are not treated as transactions; their entry during file management does not set off the replanning process. Such changes will be reflected in the inventory, requirements, and order status only after the item master record has been accessed for the next transaction entry. Special programs are needed in net change MRP to change all status data when changes in these classes are entered into item files.

Allocation of On-Hand Quantities

Maintaining interlevel equilibrium places a special demand on net change processing logic. When a planned order for a manufactured item is released, three things happen in MRP:

1. Planners select the order quantity desired (this may or may not be the same as the planned order lot size) and notify the computer MRP program.
2. MRP places the released order quantity in the completion time-bucket as a scheduled receipt (or whatever else released orders are called).
3. The required quantities of its components are shown as "allocated" in each component record.

Order-release action normally takes place when a planned order is recommended by MRP for release in the current period. Planners may override this recommendation if special conditions exist. Further discussion, including an example, may be found in Chapter 10 (p. 248) under the heading "Releasing Planned Orders."

The example in Fig. 5-10 shows how the record in Fig. 5-5 would appear at the beginning of period 2. The 25 planned order in this (now current) bucket signals a recommendation for order-release action. After the planner decides to follow this recommendation and has notified MRP, the record appears as shown in Fig. 5-11; the planned order disappears from view. If no other changes are made, gross requirements at the component level will be reduced accordingly.

This reduction would distort the status of the lower-level components *unless the required quantities of each component have been removed simultaneously from its inventory*. Usually, actual movement lags and these components are

Lead time: 3 Order quantity: 25		Period 2	3	4	5	6	7	8	9
Gross requirements		2		10	13		20	4	
Scheduled receipts			20						
On hand	4	2	22	12	-1	-1	-21	-25	-25
Planned-order releases		25							

Figure 5-10. A mature planned order.

Lead time: 3 Order quantity: 25		Period 2	3	4	5	6	7	8	9
Gross requirements		2		10	13		20	4	
Scheduled receipts			20		25				
On hand	4	2	22	12	24	24	4	0	0
Planned-order releases									

-25

Figure 5-11. Status following the posting of order release.

still in inventory, included in the on-hand data for each item as shown in Part B of Fig. 5-12. When there is such a time lag between order release and physical component disbursement (with adjustment of the inventory record), another MRP processing step is needed.

In Part A of Fig. 5-13, the problem is solved by setting up an "Allocated" field in the component item record and by posting to it the gross requirement to satisfy the parent-order being released. The quantity allocated to released parent orders, sometimes called *uncashed requisitions,* serves in lieu of the corresponding planned orders to maintain interlevel equilibrium.

Part B of Fig. 5-13 shows the next step. Following physical disbursement of the item, processing this transaction reduces both the on-hand

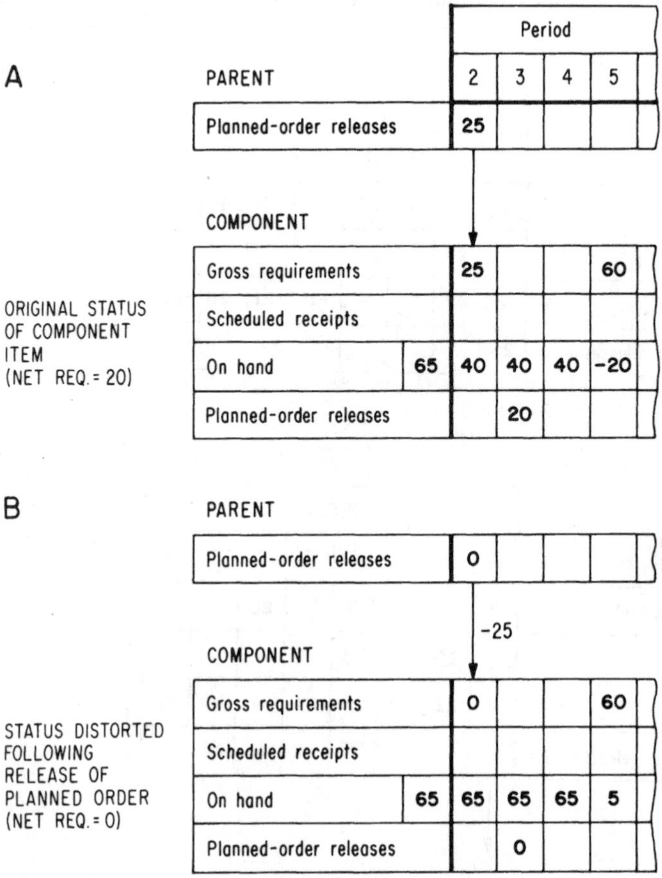

Figure 5-12. Distortion of status following order release.

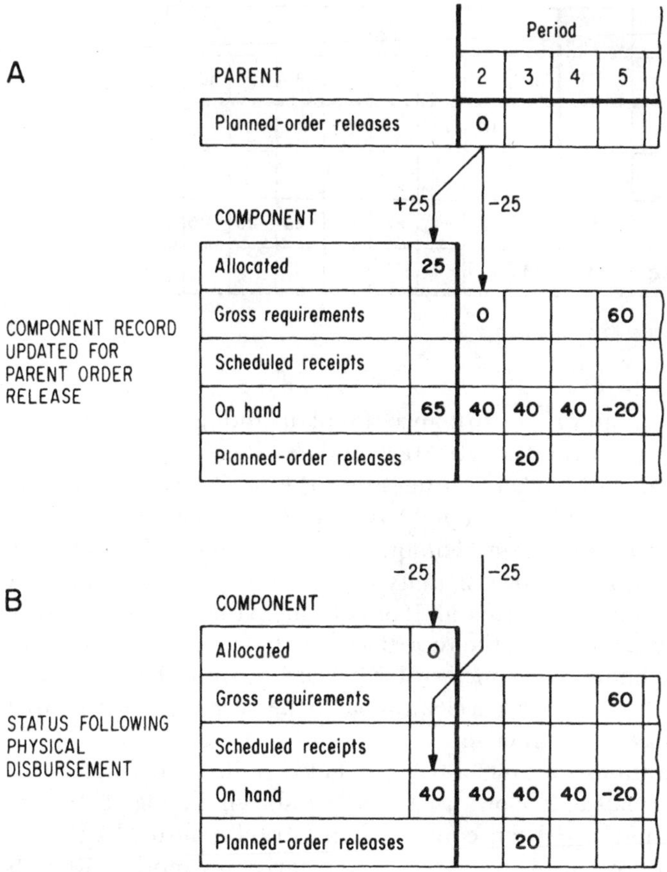

Figure 5-13. Function of allocation field.

and allocated fields by the same amount. Allocations have other important uses, which are covered in Chapter 9. These include

1. Cycle counting to detect causes of record errors
2. Testing assembly schedules for component availability

Control-Balance Fields

Another feature of net change MRP is a "control-balance" column in the time-phased inventory data, shown in Fig. 5-14. These fields provide

Lead time: 3 Order quantity: 25	Control balance	Period							
Allocated	1	3	4	5	6	7	8	9	10
Gross requirements	1		10	13		20	4		
Scheduled receipts		20							
On hand	3	22	12	-1	-1	-21	-25	-25	-25
Planned-order releases	25	25							

Figure 5-14. Delinquent performance.

information to monitor performance to plan and generate reports indicating poor performance. As MRP updates item-record files, any remaining contents of buckets representing the period just passed are dropped into the control-balance fields, except for on-hand data.

Control balances represent delinquent performance. The plan in Fig. 5-10 for period 2 shows 2 units to be consumed (disbursed or shipped) and 25 units to be ordered. If only 1 unit of this component were disbursed and none ordered, the record would appear as in Fig. 5-14. In this case, the delinquent planned-order release quantity has also been added to the period-3 bucket as a clear signal to the inventory planner that needed action has not been taken.

Negative quantities in control-balance fields indicate premature or excessive performance. Following the status shown in Fig. 5-10, for example, if 3 units had been consumed in period 1 instead of the 2 planned, 20 units on the open order received early and another 40 units placed on order, the status record would appear as in Fig. 5-15. Excessive

Lead time: 3 Order quantity: 25	Control balance	Period							
Allocated		2	3	4	5	6	7	8	9
Gross requirements	-1			10	13		20	4	
Scheduled receipts	-20			40					
On hand	21	21	21	51	38	38	18	14	14
Planned-order releases	-15								

↓ -40

Figure 5-15. Premature or excessive performance.

caution, as in this example, can do more harm than just building inventories. More on this topic may be found in Chapter 10.

Control balances representing deviations from plan may be printed and issued in each period for follow-up and corrective action. Net change programs easily generate such reports, but this is not true of regenerative programs, which lack needed data. A comparison of regenerative and net change methods of updating MRP is given in Table 5-1.

Table 5-1. Characteristics of MRP Systems

	Regenerative	Net change
Master production schedule		
Viewed as	Consecutive issues	Continuum
Input to MRP system	Entire contents	Net difference from previous status
Explosion	Full, periodic	Partial, continuous
Requirements data		
Logically integral to item record	No	Yes
Up-to-date maintenance	No	Yes
Method of generation	Reconstituted	Modified, updated
Item inventory status		
File update	Limited to inventory data	Includes inventory and requirements data
Status in narrower sense	Maintained continuously	Not separately maintained
Status in broader sense	Reestablished periodically	Maintained continuously
Records in balance	Only at explosion time	At all times
Interlevel equilibrium		
Establishment	Reestablished only at explosion time	Maintained continuously
Effect of transaction entry	Only updates record directly affected	Transaction-triggered partial explosions
Logical requirement for allocation	No	Yes
Operating phases		
Requirements planning	Periodic, long intervals	No distinction
File updating	Intracycle, short intervals	No distinction
Performance control reporting	No	Yes

Evaluating Net Change

Net change in MRP makes it possible to

1. Minimize the requirements planning job after MPS revision
2. Process schedule changes occurring between revisions
3. Keep MRP independent of the timing of revisions
4. Keep MRP continually up-to-date
5. Generate notices communicating needs for user actions

From the user's point of view, the greatest advantage of a net change MRP program is more timely response to a wide variety of changes. On the other hand, it has four disadvantages:

1. Net change lacks regeneration's self-purging capability of removing errors from files.
2. It requires stricter disciplines in operating procedures.
3. Data processing is relatively inefficient.
4. It can cause excessive nervousness from replanning.

Regeneration tends to purge errors caused by faulty data handling by MRP users. Every run, the whole master production schedule and all active bills of material and inventory data are accessed and processed; the old plan is literally thrown away, including all old errors. These errors may reappear, however, until their cause has been removed. Most companies using net change regenerate the complete program to take advantage of its purging actions periodically; the frequency decreases as performance and disciplines improve.

The net change need for stricter user discipline is a relative disadvantage only; sloppy practices and lack of data integrity will result in invalid outputs of both regeneration and net change MRP, although the former has some self-purging power. Lacking such power, net change requires more continuous and tighter disciplines in data handling, plus periodic audits to detect errors and attack their causes promptly.

Net change is a less efficient computer program than regeneration, primarily due to multiple accessing of inventory record data when handling transactions and exploding requirements. The emphasis of MRP, however, is on better inventory management and production scheduling, not on data processing efficiency.

The most challenging aspect of net change is its hypersensitivity or nervousness. File updating is equivalent to replanning, and may result in continual revision of recommended user actions, some suggested by MRP only a brief time before. The experience of almost every net change user has been frustration with the inability of people to react as quickly as computers.

In evaluating net change, two criteria should be considered:

1. MRP being up-to-date, perhaps to the minute
2. The frequency of actions needed to respond to its signals

Replanning cycles and action-notice cycles are established arbitrarily by MRP users. A full discussion of considerations in deciding on these cycles is found in Chapter 9.

A net change MRP program offers users a range of responses, from zero-delay to weekly and even monthly cycles. The relative promptness of reaction to change is a function of the type of change; net change provides many choices. The system's central architecture remains unaffected by external arrangements for data input or retrieval and by the technology of input-output devices used. It can be transformed, whenever users are ready, into an on-line, communications-oriented inventory management system without extensive reeducation or fundamental system overhaul.

6
Lot Sizing and Safety Stock

Precision without accuracy is a total waste.

Lot Sizing

Since the introduction of MRP, the problem of determining "economic order quantities" (EOQs) has shifted from square-root formulas with assumptions of average usage rates to calculations using discrete time-phased period demands. Lot sizing is the best-researched aspect of material requirements planning. Many distinct techniques are available, the most important of which are described and evaluated in this chapter.

The Concept of Economic Order Quantities

Economic order quantity (EOQ) theory applies when items are replenished in batches, not continuously, and when usage rates are low compared to replenishment batch sizes so that a significant amount of inventory is carried. How much of an item to order is the quantity that *best balances the number of orders placed over time, and the size of the orders, to yield the lowest total costs* related to the decision. The benefits from proper applications of lot-sizing techniques come from

1. Reduced expenditures associated with ordering
2. Lower charges for carrying the resulting inventory

Capturing these benefits calls for more than just changing lot sizes. Reductions in ordering costs will be achieved only if the time saved is actually used productively. Making more setups to run smaller batches producing a given total number of parts reduces work center capacity, perhaps making it inadequate to meet total production requirements. Making smaller batches will reduce total capital required (and charges incurred by it) only if total inventory is cut. This requires tight capacity control so that total input to inventory is less than total output.

Costs in Lot Sizing

Two categories of costs enter into decisions of how much of an item should be purchased or made. These are

1. *Ordering costs*—a composite of all costs related to placing purchase orders or preparing work orders, including
   a. Processing paperwork—preparing requisitions, purchase orders, receiving documents for purchased materials, and shop packets for manufactured items
   b. Changing machine and work station setups
   c. Inspection, scrap, and rework associated with setup
   d. Record keeping for work-in-process

2. *Inventory carrying costs*—the total of costs related to carrying the resulting inventory, including
   a. Obsolescence caused by market, design, or competitors' product changes
   b. Deterioration from long-term storage and handling
   c. Record keeping
   d. Taxes and insurance on inventory
   e. Storage costs for equipment, space, heat, light, and people
   f. Cost of capital invested in inventory, or foregone earnings of alternate investments

Ordering and inventory carrying costs can rarely be determined from traditional cost-accounting data; they have to be "engineered" specifically for each company's operations. While ordering costs can be estimated fairly accurately, they should be actual out-of-pocket costs, and only those costs that are affected by the decision of how many to buy or make.

Carrying cost, expressed usually as a decimal fraction of inventory value, may appear precise, but in reality will be only very approximate. Estimates of several factors will obviously be little more than educated guesses at best, particularly the last one listed above. Which of the two choices is used for this factor depends on company policy.

Table 6-1. Carrying Cost and Lot Size Relationship

Desired lot size reduction, %	Relative values: Carrying cost	Relative values: EOQ
	100	100
10	123	90
20	156	80
30	200	70

In practice, carrying cost values vary from as low as 15 to as high as 80 percent per year, and can change during a year. Higher values are used by companies that must procure outside capital rather than use retained earnings, and by those who believe that lot-sizing decisions should be charged with the same rates the business expects other capital investments to earn.

Many professionals in inventory management believe that detailed studies to estimate carrying costs are unwarranted. They prefer to view these as "management policy variables" to achieve management's objectives in inventory investment. Increasing the carrying cost used in EOQ computations will result in smaller lot sizes, and vice versa. The relationship between carrying costs and lot sizes is shown in Table 6-1. Thus the inventory carrying cost in use at any given time reflects the premium that management is putting on the conservation of capital.

Order sizing creates cycle-stock or lot-size inventory in both order point and MRP techniques. In reality, the average amount of such inventories is not equal to the theoretical one-half the quantities being ordered, as assumed in traditional EOQ calculations. In MRP, the lack of validity of such an approximation is clear. Order quantities determined by such techniques for a given inventory item will equal net requirements for one or more planning periods, causing the quantity ordered and the inventory to vary significantly from one order to the next.

The number of periods covered by an order quantity will be affected by the relative continuity of demand for the item. In cases of very intermittent demand, the order quantity often will equal the requirement for only one period. This usually will be true also for all assembled items, because of typically minor assembly setup considerations.

Lot-Sizing Techniques

The most commonly used lot-sizing techniques are

1. Fixed order quantity (FOQ)

2. Economic order quantity (EOQ)
3. Lot for lot (L4L)
4. Fixed period requirements (FPR)
5. Period order quantity (POQ)
6. Least unit cost (LUC)
7. Least total cost (LTC)
8. Part-period balancing (PPB)
9. Wagner-Whitin algorithm (WWA)

The first two are based on demand rates; the others are called *discrete lot-sizing techniques,* because they generate order quantities that equal the net requirements in an integral number of consecutive planning periods. Discrete lot sizing does not create "remnants"—unused quantities carried in inventory for some time without being sufficient to cover the next period's requirements in full.

Lot-sizing techniques can be categorized into those that generate fixed, repetitively ordered quantities and those that generate varying lot sizes. This distinction between fixed and variable is different from that between static and dynamic order quantities. A static order quantity is one that, once computed, continues unchanged over the planned-order horizon. A dynamic order quantity is continuously recomputed when required by changes in net requirements. A given lot-sizing technique can generate either static or dynamic order quantities, depending on how it is used. Of the nine techniques listed above, only the first is always static and the third always dynamic. The rest, including the EOQ, can be static or dynamic at the user's option. The last four are expressly intended for dynamic replanning.

While these techniques are more frequently used for manufactured items, the logic on which they are based is applicable to purchased materials, which have ordering and inventory carrying costs also. Any of the lot-sizing techniques can be used if appropriate modifications are made for additional factors like quantity discounts and special setup and freight charges.

Fixed Order Quantity

In practice, fixed order quantities (FOQ) are used for items with conditions not recognized by lot-sizing algorithms. These include customer order quantities for make-to-order products, limited shelf life, capacities of production equipment or processes, tool life, and unit packaging lots. FOQ are part of each item's master record and are time-phased, as shown in Fig. 6-1, to cover net requirements. "Planned order coverage" indicates due dates, not start dates.

Period	1	2	3	4	5	6	7	8	9	Total
Net requirements	35	10		40		20	5	10	30	150
Planned-order coverage	60			60					60	180

Figure 6-1. Fixed order quantity.

Unexpectedly high net requirements exceeding FOQ are signaled by MRP to planners, who make the decision of whether to increase the lot size, order two lots, or adjust parent item demand causing the high requirement. This also applies to EOQs when they are used as fixed quantities, repetitively ordered over a period of time (typically one year).

Economic Order Quantity

The EOQ formula was first derived by Ford W. Harris in 1915. It is the oldest technique in the field and is still an aid to sound planning when its limitations are recognized. Although it predated and was not intended for an MRP environment, EOQ can easily be incorporated into an MRP program. Figure 6-2 shows EOQ coverage of the same net requirements used in the previous example. These net requirements data will be carried over into subsequent examples of lot sizing, to point up the differences in the performance of the various techniques. The periods will be assumed to represent months, and the following cost data will be used throughout:

$$\text{Setup (S)} = \$100$$
$$\text{Unit cost (C)} = \$50$$
$$\text{Carrying cost (I)} = 24\% \ (0.24) \text{ per year}$$
$$\text{Carrying cost (Ip)} = 2\% \ (0.02) \text{ per period}$$

Annual usage (U units):

$$150 \text{ for 9 months demand} \times \frac{12}{9} = 200 \text{ units}$$

Period	1	2	3	4	5	6	7	8	9	Total
Net requirements	35	10		40		20	5	10	30	150
Planned-order coverage	58			58				58		174

Figure 6-2. Economic order quantity.

With these data, the cost of carrying one unit of the item in inventory for one period is $1. Using the simplest formula, the EOQ calculation is as follows:

$$Q = \text{economic order quantity} = \sqrt{\frac{2US}{IC}}$$

$$= \sqrt{\frac{2 \times 200 \times 100}{0.24 \times 50}} \qquad = \sqrt{3333} = 58$$

Many variations of this simple formula are used to handle average monthly or quarterly usage, provide *Q* in dollars, modify *Q* when lots come into stock gradually (not all at one time), for families of items using common setups, and for purchase quantity discounts. References in the Bibliography contain such details.

In this example, projected future demand from MRP, rather than forecast demand (often assumed to be equal to historical usage) is used. This illustrates the problem all forward-looking lot-sizing techniques face: a finite, or limited, planning horizon. In our example, the EOQ formula requires a year's demand data, but the system provides only nine months. Most discrete lot-sizing techniques do not require annual usage, but they assume that the total order quantity will be used up.

Limits to the effectiveness of EOQ in a discrete-demand environment are seen in Fig. 6-2. The first lot of 58 leaves a "remnant" of 13 pieces carried in inventory in periods 1 through 3 to no purpose, and 6 pieces are carried unnecessarily in periods 4 through 7 due to the excess of the second lot over requirements. Ordering three times, in quantities of 58, will be a relatively poor choice in comparison with other examples that follow.

EOQ is based on an assumption of continuous, steady-rate demand, and it performs well only where the actual demand approximates this assumption. The more discontinuous and nonuniform the demand, the less effective EOQs will be. EOQ also assumes that ordering and inventory carrying costs are the only significant ones to consider. The later section of this chapter headed "Practical Considerations in Lot Sizing" (p. 141) covers many other factors.

Lot for Lot

The L4L technique, sometimes called *discrete ordering,* is the simplest and most straightforward of all. Figure 6-3 shows an example of this method, which provides period-by-period coverage of net requirements; the planned order quantity always equals the net requirements. These order

Period	1	2	3	4	5	6	7	8	9	Total
Net requirements	35	10		40		20	5	10	30	150
Planned-order coverage	35	10		40		20	5	10	30	150

Figure 6-3. Lot-for-lot approach.

quantities are dynamic; they should be recomputed whenever net requirements change.

The use of this technique minimizes inventory carrying costs for expensive purchased or manufactured items. Theoretically, inventory is carried only for the time between receipt and use. It performs well also for items with low setup costs (assemblies) and those with highly discontinuous demand (service parts). Items in high-volume production and items that pass through specialized facilities geared to continuous production (with permanent setup) normally are also ordered lot for lot.

Fixed Period Requirements

The FPR technique is equivalent to the old rule of ordering "X months' supply" used in some stock replenishment systems, except that it determines the supply not by forecasting but by adding up discrete future planned net requirements. This method is illustrated in Fig. 6-4. Its rationale is similar to the fixed order quantity approach—the span of coverage is determined arbitrarily. With this technique, the user estimates how many future periods every planned order should cover.

Using the Fixed Order Quantity technique, the quantity is constant but ordering intervals vary; the Fixed Period Requirements technique keeps the ordering interval constant but varies the quantities. In the example shown, two periods' requirements were specified and FPR orders every other period, except when zero requirements in Period 3 extend the

Period	1	2	3	4	5	6	7	8	9	Total
Net requirements	35	10		40		20	5	10	30	150
Planned-order coverage	45			40		25		40		150

Figure 6-4. Fixed period requirements.

ordering interval. The zero requirement in period 5 is included in the two periods' requirements specified.

Period Order Quantity

The POQ technique, sometimes called Economic Time Cycle, is identical to FPR except that the ordering interval is computed using the logic of EOQ. The EOQ is computed with the standard formula, in which future demand is the MRP net requirements schedule of the item. It is then converted to the equivalent number of orders per year. The number of planning periods in a year is then divided by this quantity to determine the ordering interval.

Using the previous EOQ example,

$$\text{Orders per year} = \frac{\text{Annual Demand}}{\text{EOQ}} = \frac{200}{58} = 3.4$$

$$\text{POQ} = \frac{\text{Periods per year}}{\text{Orders per year}} = \frac{12}{3.4} = 3.5$$

(which is rounded to 3 or 4)

Figure 6-5 shows the technique applied to our example using a POQ of 3. In this, periods with zero requirement are ignored. The third order covers only one period's requirements because of the short horizon. Even with longer horizons, the last order will often be too small if demand continues beyond. Some practitioners attempt to extend the horizon arbitrarily by using average period demand or more sophisticated calculations based on the previous net requirements patterns. A much better solution is to shorten cycle times so that *no released order of any component below this level is affected by this last order.*

Both fixed-interval techniques, FPR and POQ, avoid remnants and thus reduce inventory carrying costs. For this reason they are more effective than EOQ (for the same number of periods), since setup cost per year is the same but carrying costs will be lower. Periodic ordering techniques are simple, avoid remnants, generate orders at regular inter-

Period	1	2	3	4	5	6	7	8	9	Total
Net requirements	35	10		40		20	5	10	30	150
Planned-order coverage	85					35			30	150

Figure 6-5. Period order quantity.

vals, and help smooth work input to gateway (starting) work centers. In comparison with other discrete lot-sizing techniques to be described, FPR and POQ—like that of classic EOQ on which they are based—are handicapped by discontinuous, nonuniform demand.

Least Unit Cost

The LUC technique and the three that follow have certain things in common. All allow both the lot size and the ordering interval to vary. They share an assumption that a portion of each order, equal to the quantity of net requirements in the first period covered, is consumed immediately upon arrival in stock and thus incurs no inventory carrying charge. Inventory carrying cost, under all four of these lot-sizing methods, is computed on the basis of this assumption rather than on average inventories in each period. All four of the techniques share the EOQ objective of minimizing the sum of setup and inventory carrying costs, but each employs a different calculation.

LUC is an iterative trial-and-error approach, determining the order quantity by asking whether it should equal only the first period's net requirements or should be increased to cover the next period's requirement, and the one after that, and so on. A "unit cost" is calculated for each step by dividing the total of setup and carrying costs by the cumulative lot quantity at that step. The final decision is based on the lowest unit cost.

Table 6-2 contains the computation for the first lot. The next one is computed the same way, starting with period 4. As Fig. 6-6 shows, the least unit costs are found for

1. A lot quantity of 45, which covers periods 1 and 2
2. An order of 60, which covers periods 4 through 6

Table 6-2. Computation of Least Unit Cost

Setup: $100; Inventory carrying cost: $1 per unit per period

Period	Net require ments	Carried in inventory (periods)	Prospective lot size	Carrying cost, $ For lot	Carrying cost, $ Per unit	Setup per unit, $	Unit cost, $
1	35	0	35	0	0	2.86	2.86
2	10	1	45	10.00	.22	2.22	2.44
3	0	2					
4	40	3	85	130.00	1.53	1.18	2.71

Period	1	2	3	4	5	6	7	8	9	Total
Net requirements	35	10		40		20	5	10	30	150
Planned-order coverage	45			60			45			150

Figure 6-6. Least unit cost.

3. A third order of 45, covering periods 7 through 9

The unit cost varies, often widely, from one lot to the next. Tradeoffs between consecutive lots could reduce the total cost of two or more lots. Our example shows this. If the requirement (5) in period 7 were added to the quantity of the second lot (60), its inventory carrying cost would increase by \$15, but that of the next lot would decrease by \$40. The lot-sizing technique described next attempts to overcome this flaw in LUC logic. Many simulations have shown that LUC produces results less economical than those of other techniques. Since it requires more calculations than these, it has dropped from practical application.

Least Total Cost

The LTC technique is based on the premise that the sum of setup and inventory carrying costs ("total cost") for all lots within the planning horizon will be minimized if these costs are equal; this is the same as classic EOQ. The LTC technique attempts to reach this objective by ordering lot quantities for which the setup cost per piece and the carrying cost per piece are most nearly equal. As Table 6-2 shows, LUC didn't do this; it chose a quantity at which setup cost per piece (\$2.22) significantly exceeds carrying cost per piece (\$0.22).

LTC involves a series of iterations, comparing ordering and carrying costs for a succession of increasingly larger lots as shown in Fig. 6-7. If the lot size were equal to the first period's requirement (35), no inventory would be carried (an assumption of LTC) and the total cost would be equal to the \$100 setup. Adding the second week's requirement of 10 would incur \$10 (= 0.02 × 50 × 10) cost for carrying the inventory but would avoid another \$100 setup—thus giving a lower total cost.

Continuing the process, the zero requirement in Period 3 adds nothing to carrying cost. However, adding the fourth month's requirement (40) increases the total carrying cost to \$130, more than the cost of another setup. This is the least total cost, where both costs are "most nearly" equal. Obviously, adding another requirement quantity to the lot size would

Item cost: $50; carrying cost: 2% per period

Period	Net require- ments	Cumu- lative lot size	Excess inven- tory	Periods carried	Carrying cost This lot	Carrying cost Cumu- lative	Ordering cost$	Total cost $
1	35	35	0	0	0	0	100	100
2	10	45	10	1	10	10	100	110
3	0	45	0	0	0	10	100	110
4	40	85	40	3	120	130	100	230
5	0	85	0	0	0	130	100	230
6	20	105	20	5	100	230	100	330

Recommended lot size = 85

Figure 6-7. Least total cost.

only widen the difference between carrying and setup costs. The recommended lot size is 85 pieces.

LTC is favored over LUC, but its logic has a flaw of its own in the premise that the lowest total cost results when the inventory carrying cost and setup cost are (most nearly) equal. This is theoretically true for EOQ but not for discrete lot-sizing approaches, which assume that the portion of the lot covering the first net requirement incurs no carrying cost.

Figure 6-8 graphs the cost relationships, with $100 setup cost, for a series of uniform discrete demands:

Period:	1	2	3	4	5	6	7
Demand:	20	0	20	0	20	0	20

In the classic EOQ model, the total of ordering and carrying costs is a minimum at the point of intersection of the carrying cost line and the ordering curve (where they are equal), only when the line passes through the intersection of X (order quantity) and Y (cost) axes, as it would for data in Fig. 6-7. In discrete lot-sizing models, however, the carrying cost line crosses the Y axis below zero because of the assumption regarding the first usage.

In a graph of nonuniform demand, there would be not one line but several connected shorter lines, each with a different slope. This is why simulations of LUC reveal that the ratio of the setup and carrying cost elements is lopsided, sometimes one way and sometimes the other. In most cases, however, the LUC ordering cost will be larger than the

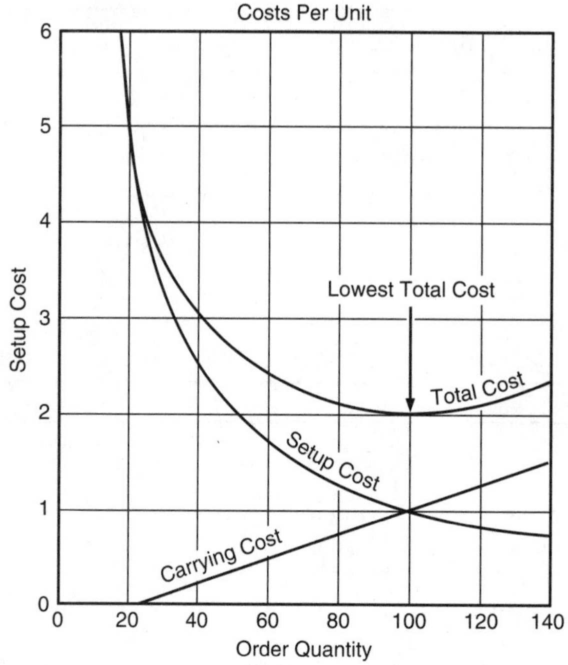

Figure 6-8. Cost relationships for discrete-demand series.

carrying cost. The LTC technique, which seeks to equalize these elements, is therefore biased toward larger order quantities.

Part-Period Balancing

The PPB technique employs the same logic as LTC but its computations are simpler, avoiding the laborious iterative computations of LTC by calculating an "economic part-period factor" (EPP). A *part-period* is one

Table 6-3. Computation of Least Total Cost

Period	Net requirements	Carried in inventory (periods)	Prospective lot size	Part-periods (cumulative)
1	35	0	35	0
2	10	1	45	10
3	0	2		
4	40	3	85	130

unit of an item carried in inventory for one period; it is a convenient expression of inventory carrying cost for purposes of comparison and tradeoff. Table 6-3 shows the calculations for the data in our example.

The Economic Part-Period factor is defined as *that quantity of an item which, if carried in inventory for one period, would result in a carrying cost equal to the cost of setup*. It is computed by dividing the setup cost per lot (S) by the inventory carrying charge per unit per period (IpC). In our example,

$$\text{EPP} = \frac{S}{\text{IpC}} = \frac{100}{0.02} \times 50 = 100$$

The PPB technique selects that order quantity at which the part-period cost most nearly equals the EPP. Figure 6-8 shows the technique applied to our example; the quantity chosen for the first lot is 85, because its 130 part-periods most nearly approximate the EPP of 100. This order would cover requirements of periods 1 through 5. A second order of 65 would cover total requirements of periods 6 through 9.

An adjustment routine included in PPB, called *Look-ahead/Look-back*, is intended to prevent stock covering peak requirements from being carried for long periods of time and to keep orders from being brought in too early in periods with very low requirements. The adjustments are made only when the conditions exist that Look-ahead/Look-back corrects; in other cases, PPB and LTC will yield identical results. This would be the case with the demand data used in the previous examples.

To demonstrate Look-ahead/Look-back, a different series of net requirements will be used:

Period:	1	2	3	4	5	6	7	8	9
Net Reqts:	20	40	30	10	40	30	35	20	40
Plnd Ord:			X→				Y→		

With an EPP of 100, a first lot of $X = 90$ is selected. This covers periods 1 through 3; the next lot Y would be sized for later periods. But before this is firmed up, the technique looks ahead to period 5, sees that its 40 units would have to be carried in inventory for one period, which would cost 40 part-periods. If, on the other hand, the 10 units in period 4 were added to the 90 in the first lot, they would be carried for 3 periods at a cost of only 30 part-periods, a more economical result. The look-ahead test is repeated for successive pairs of period demands until it fails. The complete planned-order schedule, adjusted for look-ahead, is shown in Fig. 6-9.

Period		1	2	3	4	5	6	7	8	9	Total
Net requirements		20	40	30	10	40	30	35	20	40	265
Coverage	without look-ahead	90			80			95			265
	with look-ahead	100				105			60		265

Figure 6-9. Part-period balancing with look-ahead.

To prevent the look-ahead feature from trying to overcome a steep upward trend in the demand (creating very large order quantities and defeating the logic of least total cost), an additional test is made. The part-period cost of the last period-demand covered by the prospective lot is compared with the EPP, and the look-ahead process is stopped if this cost equals or exceeds the EPP.

The look-ahead test is always made first. If covering an additional period is uneconomical, the look-back test is made. This checks the desirability of cutting the lot size, adding the requirements in the last period covered by the order to the next lot. Consider the following net requirements schedule:

Period:	1	2	3	4	5	6	7	8	9
Net Reqts:	20	40	30	15	30	25	50	20	40
Plnd Ord:			←X			←Y			

Again the first lot is 90 to cover periods 1 through 3, with the next lot covering period 4 and beyond. Looking back to period 3, however, this incurs 60 part-periods (to carry these 30 pieces 2 periods), whereas it would cost only 15 part-periods (to carry the period-4 requirement of 15 for 1 period in the second lot) if the first lot covered only periods 1 and 2. The complete planned-order schedule after look-back appears in Fig. 6-10.

Period		1	2	3	4	5	6	7	8	9	Total
Net requirements		20	40	30	15	30	25	50	20	40	270
Coverage	without look-back	90			70			110			270
	with look-back	60		75			95			40	270

Figure 6-10. Part-period balancing with look-back.

It might seem that the Look-ahead/Look-back features of PPB improve its effectiveness. Before concluding this, however, study the look-ahead example of Fig. 6-9. Obviously it is cheaper to carry 10 units for three periods than 40 units for one period, but there are other consequences. The 30 in period 6 also will be carried for one period less, saving 30 part-periods. The 35 in period 7, however, would not have incurred any carrying cost had look-ahead not been employed; now it will cost 70 part-periods. The look-ahead feature of part-period balancing does not look ahead far enough. When the adjustment is made, it changes the timing and coverage of all subsequent planned orders in the schedule, with results that the technique does not evaluate.

In our example, look-ahead saves a total of 130 part-periods and incurs added costs of 100 part-periods. The last lot of 60, however, incurs only 40 part-periods due to lack of horizon; it eventually will be recomputed and increased to get closer to the 100 EPP. This will undoubtedly more than offset the net saving of 30 part-periods in periods 1 through 9. By then, however, the basis will have been lost for making a valid comparison of the alternative strategies.

The look-back proposition suffers from the same shortcoming; it fails to examine the consequences of the adjustment throughout the planning horizon. In our example in Fig. 6-10, look-back produces a net saving of 75 part-periods but adds a fourth setup costing the equivalent of 100 part-periods. If setup were larger relative to unit cost, the EPP would be higher and the lots spaced further apart. The last period-demand covered by any prospective lot would then almost always entail more part periods than the first period of the subsequent lot and look-back would operate more frequently. This would result in more smaller orders, which would subvert the logic of the least total cost technique on which look-back is grafted.

There is another important consideration. MPS changes may affect many components, generating the nervousness problem discussed in Chapter 10. Look-ahead/Look-back will multiply these problems, affecting both quantities and timing of orders.

Wagner-Whitin Algorithm

This technique attempts to determine an optimum lot size by evaluating all possible order quantities to cover net requirements over the total planning horizon. The mathematics of WWA is "elegant"; it reaches this objective without actually having to consider, specifically, each possible strategy. Its solution to lot sizes for the net requirements schedule used in previous examples is shown in Fig. 6-11.

WWA minimizes the combined (total) costs of setup and of carrying inventory over the total planning horizon. It can be used as a standard for

Period	1	2	3	4	5	6	7	8	9	Total
Net requirements	35	10		40		20	5	10	30	150
Planned-order coverage	45			65				40		150

Figure 6-11. Wagner-Whitin algorithm.

measuring the relative effectiveness of the other discrete lot-sizing techniques. Its disadvantages are a high computational burden, nervousness, and the difficulty nonmathematicians have understanding its mechanics.

Computation time is not significant with current computer technology, taking only microseconds. The second disadvantage, however, is serious. There are typically tens of thousands of inventory items in an MRP program for which planned orders have to be computed, and requirements for many of these change frequently. Computers can easily handle such changes; suppliers and manufacturing plants cannot. The third disadvantage is overwhelming. If users can't understand how a technique works, they won't use it. Wagner-Whitin's algorithm is not adopted in practice.

Lot Size Adjustments

The planned order quantity determined by any of the lot-sizing techniques is subject to certain practical constraints; among these are the following:

1. Minimums and maximums
2. Scrap allowances
3. Multiples
4. Raw material cutting factors

Minimums and Maximums

A lot size calculated by any technique should be adjusted if the quantity to be ordered is impractical. One type of minimum already has been mentioned—the order is less than the first net requirements and must be made at least equal to it. Another is the smallest order quantity a supplier will accept. Maximums frequently are imposed by management policies. Minimums and maximums usually are stated in quantities—"no less than 50 and no more than 400 units"—but the limits may be stated in period coverage—"no less than 4 weeks and no more than 12 weeks," or "not to exceed one year's supply."

Scrap Allowances

A *scrap allowance*, or *shrinkage factor*, is a quantity added to the computed lot size to compensate for anticipated scrap or other loss in process, and to ensure that the required quantity of "good" pieces is received. This is important only in instances of discrete lot sizing, because the order quantity equals net requirements in an integral number of periods (no "remnants"). The scrap allowance normally will vary from item to item, being based usually on past experience with scrap, and is at best a guess.

The scrap allowance may be stated either as a number of pieces or as a percentage of the order quantity. A fixed percentage generally is undesirable if the order quantities vary significantly form lot to lot. In machining, for example, scrap tends to be a function of the number of different setups required in processing the lot rather than the quantity being run. Some MRP users apply a "declining percentage" formula:

$$Q = L + a\sqrt{L}$$

where Q = order quantity
L = lot size
a = multiplier reflecting scrap incidence

The value of the multiplier is chosen by planners based on past experience. With a multiplier of 1, Table 6-4 contains scrap allowances for a wide range of computed lot sizes.

The proper way to handle scrap allowances in the time-phased MRP record is to add them to (include them in) the planned-order quantities. When the planned orders are released, the full quantity, including the

Table 6-4. "Declining Percentage" Scrap Allowance

Computed lot size	Scrap allowance	Order quantity	Percentage allowed
1	1	2	100
4	2	6	50
9	3	12	33
16	4	20	25
25	5	30	20
100	10	110	10
400	20	420	5
10,000	100	10,100	1

scrap allowance, should be considered to be on order. This quantity would then be reduced as and if scrap losses were posted to the record.

The practice of including scrap allowance quantities in the item's gross requirements (in order to display the projected on-hand quantities as they are expected to be after scrap) is poor; it distorts the fundamental relationship of parent planned-order quantity to the component item's gross-requirements quantity. Furthermore, whether scrap actually will occur is uncertain. Until it does, MRP should project the item's status as though scrap will not occur.

Multiples

Another constraint on lot sizing is the requirement that a given item be ordered in multiples of some number. This may be dictated by process equipment considerations such as the number of pieces constituting a full grinder load, steel bars fed to a bar lathe simultaneously, packaging lots, and capacity of plating or heat-treating equipment. The calculated lot size, in these cases, is increased to the nearest multiple specified.

Raw Material Cutting Factors

Raw material cutting and the waste of kerfs often require adjustments to lot sizes. The lot-sizing technique, unaware of the form in which the item's raw material comes, may set an odd quantity or understate the total raw material quantity. For example, if a sheet of metal can be cut into 9 pieces to make an item, a lot size of 30 of the item will result either in a remnant of raw material left over or, more likely, the operator overrunning the order by cutting 4 sheets into 36 pieces. Making a few more pieces may be a better alternative than having a remnant to keep track of and use up somehow later.

In cases where more than one type of adjustment is to be made to the order quantity for a given item, the several adjustments should be made consecutively in a logical sequence. For example, if the unadjusted lot size is 173, equivalent to five periods' requirements, but the item is subject to a three-period ceiling, 9 percent scrap allowance, and raw material cutting constraint of 20 pieces per unit of raw material, the adjustment would be:

Unadjusted lot size	= 173
3-period ceiling	= 121
Plus 9 percent scrap allowance	= 11
To round to nearest multiple of 20, add	8
Adjusted total lot size	= 140

Minimums and maximums, scrap allowances, multiples, and raw material cutting factors may result in excess inventory. This excess, however, is temporary. It will be applied by MRP against later gross requirements and will not accumulate.

Evaluating Lot-Sizing Techniques

None of the lot-sizing techniques is perfect—each suffers from some deficiency. The difficulty in evaluating the relative effectiveness of these techniques is that their performance varies, depending on the patterns of net requirements data and on the ratio of setup and unit costs. Furthermore, some of the techniques assume gradual, steady-rate inventory depletion, whereas others assume discrete depletion, which affects the way inventory carrying costs are computed for purposes of comparison.

Ignoring these distinctions and basing inventory carrying costs on discrete depletion at the beginning of each period, the performance of the lot-sizing techniques for the data set used in the previous examples is shown in Table 6-5.

These figures are meaningful only for the net requirements schedule, the setup cost ($100), and the unit cost ($50) used in the examples. A change in these data will produce a different sequence. For example, if setup were $300, the Period Order Quantity would outperform Least Total Cost and match Least Unit Cost in effectiveness. By changing requirements data, the example can be rigged to produce any result desired, including Economic Order Quantity equal in performance to Wagner-Whitin. The factors that affect the performance of lot-sizing techniques are

1. Variability of period demands
2. Length of the planning horizon

Table 6-5. Comparison of Lot-Sizing Algorithm Performance

Algorithm	Number of setups	Setup cost, $	Part-periods	Carrying cost, $	Total cost, $
W-W	3	300	95	95	395
LUC	3	300	120	120	420
LTC	2	200	245	245	445
POQ	3	300	155	155	455
EOQ	3	300	206	206	506

3. Size of the planning period
4. Ratio of ordering and unit costs

Variability of demand includes both nonuniformity (varying size) and discontinuity (gaps of no period-demand). Short planning horizons allow special conditions to favor one technique over another; long horizons provide better comparisons. Shorter planning periods result in smaller period requirements, enabling lot-sizing techniques to get closer to the best balance between ordering and carrying costs. The ordering-cost/unit-cost ratio directly affects the frequency of ordering and thus the lot size.

There is no one "best" lot-sizing technique for a given manufacturing environment, for one class of items, or even for a single specific item. Experience with MRP has shown that the lot-for-lot approach gives good results where setup costs are small; where these costs are high, as in the fabrication of components, Least Total Cost, Part Period Balancing, or Period Order Quantity can provide satisfactory results.

In selecting lot-sizing techniques for use with MRP, detailed studies and extensive debates are not warranted; in practice, one discrete technique is about as good as another. Two exceptions are Least Unit Cost, requiring more calculations for poorer results, and Part Period Balancing, where Look-ahead/Look-back causes excessive nervousness.

Good rules of thumb for a first cut at selecting lot-sizing techniques are

1. Use Fixed Quantity at the MPS level, limiting changes to timing only, and damping them with firm planned orders.
2. At intermediate BoM levels
   a. Where setup costs are low, use Lot for Lot.
   b. Where setup costs are significant, use Period Order Quantity or Least Total Cost without Look-ahead/Look-back.
3. At bottom BoM levels (purchased items), use Period Order Quantity.

These selections minimize nervousness at very little potential cost. The dynamic techniques will capture the bulk of the savings. POQ for purchased items will fit regular supplier deliveries and help minimize freight charges. Experience with these may indicate possible savings by changing to another technique, but change only when those savings are significant.

In the final analysis, it does not matter how elaborately and with what precision lot sizes can be computed. All techniques are vulnerable to changing demand and, unfortunately, this is found in most MRP environments. The best way to ensure maximum economy in ordering materials is to develop and maintain stable plans for procurement and fabrication.

Practical Considerations in Lot Sizing

The use of formulas to calculate "economic" order quantities poses several significant problems. The assumptions made in the formula derivations that inventory carrying costs and ordering costs vary uniformly with lot size are not generally valid. These costs cannot be assigned a specific, constant value over a range of lot sizes; this value will vary with the total inventory. Either too large an increase in inventory or too many orders being generated can result from application of the formulas.

Lot-sizing techniques consider only a few of the factors involved in the decision of how much to order. Several other factors override those considered in the techniques, dictating different lot sizes. These include short material shelf life, risk of obsolescence, material availability, the item's bulk versus value, and tool life. The techniques lack ways to evaluate the benefits of not making items too soon, tying up possibly scarce materials and utilizing limited capacity on the wrong priorities.

Dynamic order quantities are a mixed blessing in an MRP environment. While they reflect the most up-to-date version of the materials plan, they change frequently the item's component requirements and thus also their planned coverage. A recomputation of a parent planned order quantity will often cause MRP to reschedule component item released orders, in addition to revising planned order quantities and schedules in future periods.

Upsetting previous decisions on component item orders can cause severe operations problems, some violating common sense. Figure 6-12 is an MRP profile of a parent and one of its components. The Period Order Quantity technique is used, and a lot size for the parent is set at 3 weeks. The component, with only this one parent, uses Lot-for-Lot. A replenishment order (30) for this component has just been released to the plant.

Shortly after, a customer order for 1 unit of the parent is canceled, reducing its Week 3 requirement from 2 to 1, eliminating the 1-unit net requirement. POQ recalculates the parent lot sizes and reschedules the planned orders as shown by the changes in Fig. 6-13. Since the larger lot (31) of the first planned parent order now exceeds the component's available 15 units, MRP signals the need to complete the component order just released in half of the normal lead time. A customer cancellation calls for expediting a component order! This is good theory run amok.

While some nervousness inevitably arises in normal operations, it can be greatly amplified by lot-sizing techniques recomputing lot sizes automatically. This can and should be avoided. Many users of MRP systems "freeze" quantities of planned orders (see "Firm Planned Orders" on p. 65 in Chapter 3) within the span of the cumulative product lead time, so that

P.O.Q. = 3 Wks. L.T. = 2 Wks.

Parent	Late	Period 1	2	3	4	5	6	7	8
Projected Requirements		2	5	2	4	6	21	4	7
Scheduled Receipts									
On Hand	8	6	1						
Net Requirements				1	4	6	21	4	7
Planned Order Receipt				11			32		
Planned Order Release		11			32				

Lot for Lot L.T. = 4 Wks.

Component	Late	Period 1	2	3	4	5	6	7	8
Projected Requirements		11			32				
Scheduled Receipts					30				
On Hand	15	4	4	4	2	2	2	2	2
Net Requirements									
Planned Order Receipt									
Planned Order Release									

Figure 6-12. Period order quantity example.

these orders cannot change gross requirements on lower levels that may be covered by replenishment orders already released.

The *aggregate effects* of changing lot sizes by introducing new techniques or using different parameters should be evaluated before these are introduced. New lot sizes may involve making an excessive total number of setups in a work area or on a machine, reducing capacity below that required to produce the required total output. On the other hand, small lot sizes may "starve" equipment like heat-treating furnaces and baking ovens, tempting people to delay processing orders through them until full loads are available. Techniques assume that adequate capital will be available to cover the total inventory generated; this may be more money than management desires for this type of investment.

P.O.Q. = 3 Wks. L.T. = 2 Wks.

		Period							
Parent	Late	1	2	3	4	5	6	7	8
Projected Requirements		2	5	1	4	6	21	4	7
Scheduled Receipts									
On Hand	8	6	1						
Net Requirements				0	4	6	21	4	7
Planned Order Receipt					31				
Planned Order Release			31						

Lot for Lot L.T. = 4 Wks.

		Period							
Component	Late	1	2	3	4	5	6	7	8
Projected Requirements			31						
Scheduled Receipts			←	—	30				
On Hand	15	15	−16	−16	14				
Net Requirements									
Planned Order Receipt									
Planned Order Release									

Figure 6-13. Revised period order quantity example.

Techniques work on one item at a time, ignoring the desirability of scheduling together families of items having similar setups. These links can be added to MRP programs using "run with" codes in item master records, calling the family of items to planners' attention when one is triggered for ordering. Adjusting lot sizes for these items to have "equal runout time" helps ensure that all are at or near reorder condition about the same time.

Planning larger lot sizes will achieve no economies if constant shortages and crises cause many of the lots to be "split" into two or more sublots. Techniques are valuable, and sound planning is possible, only where plans are well executed.

Safety Stock

Safety stock , also called *reserve* and *buffer stock,* is a cushion of inventory in excess of planned requirements to help meet unplanned needs. Order point stock-replenishment systems must include safety stock because of the fallacies in their basic assumptions—that demands during replenishment periods are uniform and can be predicted accurately.

The list of contingencies against which people desire safety stock protection is incredibly long. It includes unexpected customer demand, late supplier deliveries, machine breakdown, tool trouble, power failures, scrap and rework, record errors, and many more surprises experienced in production. There are four basic approaches to determining how much safety stock is needed on products and components: guesstimates, rules of thumb, statistical analyses, and inflating master production schedules.

Guesstimates are probably the most frequently used, being easiest to apply, and are based on planners' personal judgment. They usually increase immediately after a shortage occurs but are rarely decreased. *Rules of thumb* are equally irrational and require additional work to apply. A popular one bases safety stock on A-B-C inventory classification; expensive A-items should have little, moderate B-items some more, and low-cost C-items plenty. This ignores the protection furnished by lot sizes in excess of immediate requirements; C-items usually have very large order quantities and short replenishment lead times; they may not even need safety stock. Conversely, A-items are exposed more frequently to stock-outs because of frequent ordering and may need large safety stocks.

Statistical analyses attempt to relate the amount of safety stock planned to a desired level of "customer service," based on the variability of demand experienced in the past. A variety of techniques is covered extensively in the literature included in the Bibliography. Computer software often provides programs for including one or more of these in inventory planning.

If safety stocks are planned in MRP programs, *inflating the master production schedule* is the best means, but it is dangerous. It results in overstating the requirements for all resources unless precise adjustments are made to those in which cushions are not desired. MPS should be inflated only in the first period, carried forward each time MPS are revised. It has the advantage of planning safety stocks in balanced sets matching desired parent item plans.

Cushions can be added to plans to help guard against unexpected shortages in two other ways: safety time and safety capacity. *Safety time* is used to schedule orders earlier than otherwise required. This results in varying amounts of safety stock, depending on items' time-phased requirements. The amount of safety time usually is guesstimated or set by rules of thumb.

Safety capacity planning attempts to provide extra capacity to handle unplanned needs for additional components. It assumes that raw materials will be available to make the components needed; this is not unreasonable, since raw materials typically are ordered in large lots. When shortages occur, specific parts needed can be made quickly if capacity and materials are available. This approach has several important advantages:

1. It avoids committing common materials to specific items long before they are actually needed.
2. It requires only a fraction of the work needed for setting safety stocks, since few work centers and not many products and components are involved.
3. It is easy to monitor, using input/output capacity control to make needed adjustments to planned safety capacity.
4. It allows MRP to show truer expected capacity needs.
5. It does not distort MRP order-release or due dates.
6. It gives results superior to safety stock or safety time.

Chapter 4 shows how MRP calculations handle safety stock and safety time planning, and also discusses how planning safety stock distorts net requirements and true need dates.

The effects of safety stock planned at component levels will be clear in an example using an item with a planned lead time of 4 periods, a demand-during-lead-time forecast of 40 units, and 20 units of safety stock; hence, order point equals 60. Order point is reached, and Fig. 6-14 shows weekly inventory projections and also the period the replenishment order is due.

Period	1	2	3	4
Forecast	10	10	10	10

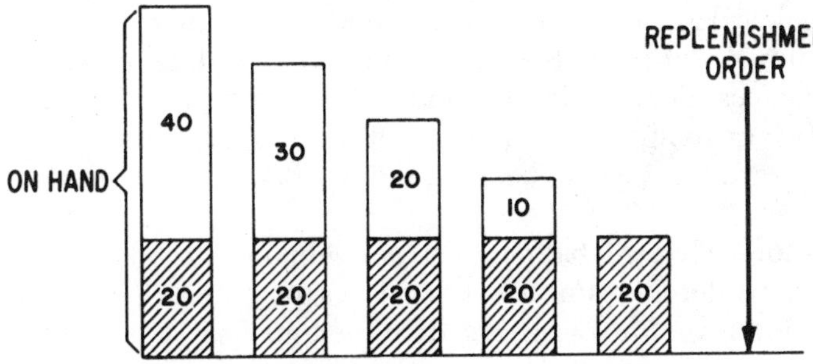

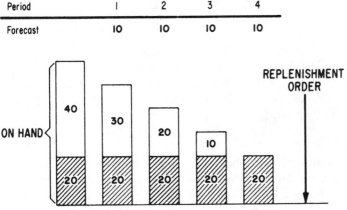

Figure 6-14. Implications of safety stock.

Planned lead time: 4
Safety stock: 20

Period	1	2	3	4
Forecast	10	10	10	10
Actual	20			

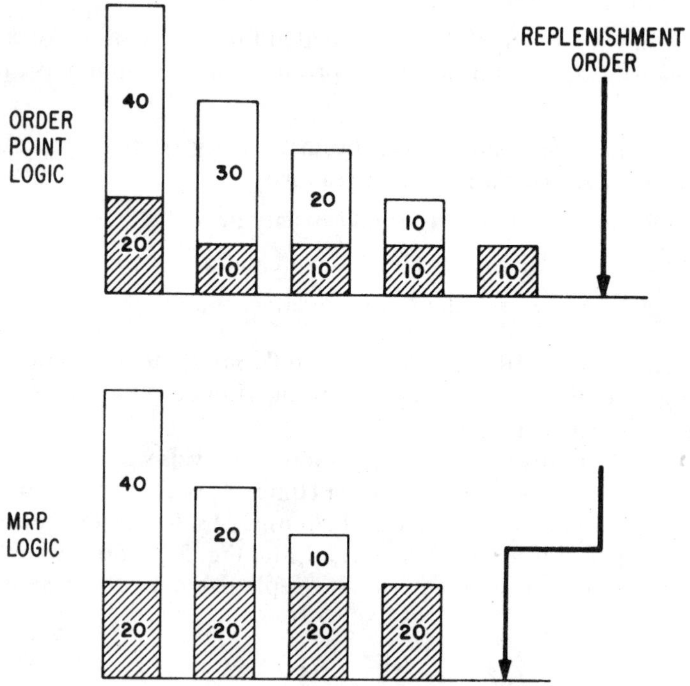

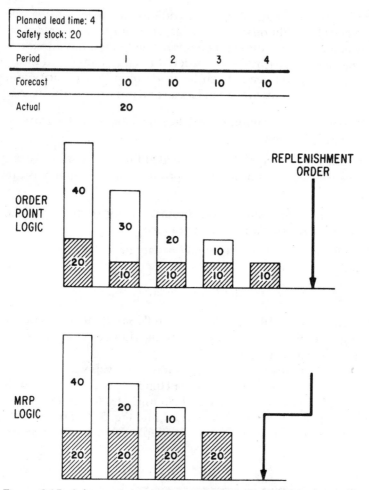

Figure 6-15. Safety stock demand exceeds forecast in first period.

As shown in Fig. 6-15, actual demand in Period 1 is 20, not the forecast 10. Using statistical order point, the timing of the replenishment order is unaffected; the excess demand has been met by using half of the safety stock. Using MRP or time-phased order point logic, the due date of the order moves forward one period, keeping safety stock at the original 20 units.

Figure 6-16 shows what will happen if actual demand exceeds forecast again in Period 2. Under OP, safety stock furnishes the excess 10 units and is now used up, but again the order due date remains firm. Under MRP or TPOP, the replenishment order is rescheduled again and planned

Planned lead time: 4
Safety stock: 20

Period	1	2	3	4
Forecast	10	10	10	10
Actual	20	20		

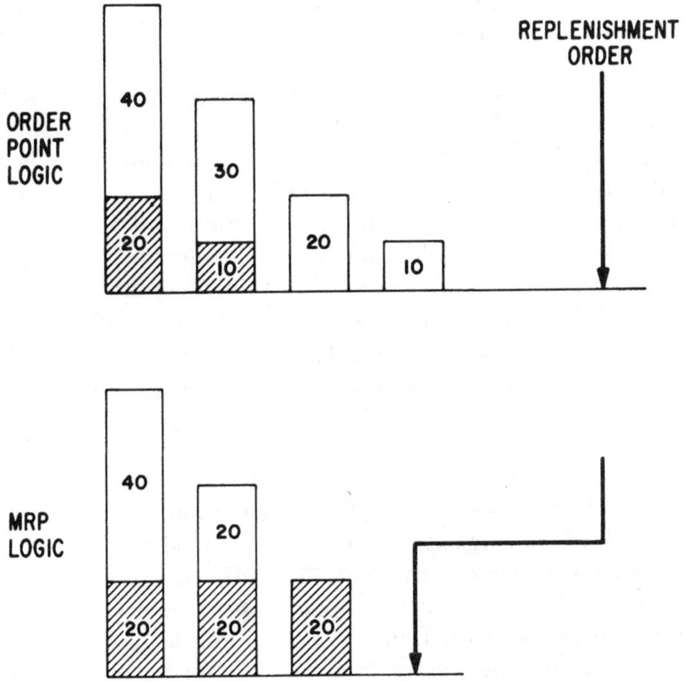

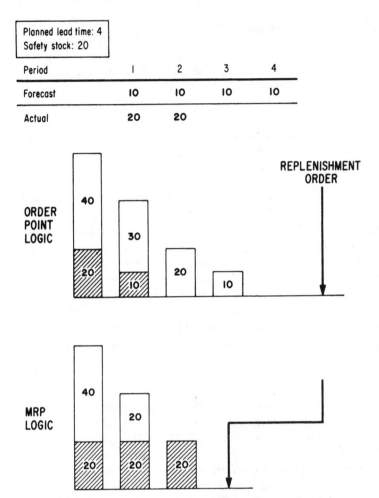

Figure 6-16. Safety stock demand exceeds forecast in second period.

safety stock remains 20 units. Now if demand in periods 3 and 4 is as forecast, all inventory under OP will be used up; this will not be true (if the replenishment order is completed on time) under MRP or TPOP. *These techniques strive to preserve safety stock intact*; if they succeed, safety stock is "dead" inventory that could be eliminated.

Besides creating "dead" inventory, safety stock in MRP has other harmful effects. People who know that this "cushion" exists may relax efforts to get orders completed on time or lose confidence in MRP when they see that order due dates are not valid. Chapter 11 contains a further discussion of this.

Practical Considerations with Safety Stocks

There are myriad reasons for wanting safety stock cushions, but there are only four ways a specific amount can be determined. Deciding which of these four to use on each of thousands of items is completely impractical.

Besides, safety cushions hinder more then help. Experience has shown that upsets rarely occur on items having safety stocks; the excess inventory is unused. When most needed, they are not there; planning efforts are wasted. Even the laws of chance seem to work against good inventory control.

When most needed, safety stocks are most difficult to get. When demand rises, safety stocks are depleted quickly, a clear indication to naive, underqualified users that increased quantities should be planned. When safety stock levels are raised, replenishment orders are released earlier, capacity requirements increase, work-in-process rises, and lead times lengthen and get more erratic, resulting in more shortages and indicating still inadequate safety stocks. Another form of the "vicious cycle" discussed in Chapter 11, this is just a rational way to get into deeper trouble.

The correct reaction to inability to get safety stocks is completely counter-intuitive: cut the planned safety stocks. This delays order release, reduces work-in-process, shortens lead times, and improves delivery performance to plan.

The implicit belief among production people that problems causing upsets are unsolvable led to their conviction that their only hope of avoiding shortages lay in having extra materials. Clear evidence that most often such cushions were simply useless, excess inventory resulted in redoubled efforts to find more sophisticated ways to plan them. Computer-based MRP tempted many to try frequent replanning to handle crises better; this also failed, because people and manufacturing plants were unable to react as flexibly as computers. Now the only real solution is becoming widely recognized and applied: *eliminate the problems.*

7
Data Requirements and Management

Swamped with data; starving for information.

Files and the Database

In a computer-based MRP program, files constitute the foundation on which the MRP structure is built. Foundations determine the soundness and utility of any structure, and file data quality underlies effective operation of MRP. Quality includes accuracy, timeliness, and accessibility of file-record data. Lack of file data integrity is a primary reason why many MRP programs installed in industry have failed to live up to expectations.

Computer programs operate on two classes of data: data input by transactions and data in files. Terms such as *data control, data management,* and *database* refer to file data. Databases consist of one or more data files, and activities relating to them are known as *file management, data management,* and *database management.* Databases, also called *data banks,* usually serve in common several applications and subsystems.

Before computer-based integrated planning and control systems became available, departments and functional groups in manufacturing companies maintained their own manual files, organized to suit their particular needs. Such files included a considerable amount of data duplication. For example, a typical inventory record contained

1. Part number
2. Part description

3. Standard cost
4. Raw material used (for manufactured parts)
5. Purchase-order quantity and quantity discounts
6. Where-used information

All of these were duplicated in at least one other, and sometimes all, of the following files:

1. Cost sheets
2. Purchasing records
3. Routings
4. Engineering drawings
5. Service part price (billing) records
6. Stores location records

Computer databases avoid such duplication. File design has two objectives:

1. Eliminating or minimizing data duplication, redundancy, different file "versions," and different update cycles
2. Optimizing file access and economy of use

These are conflicting objectives. If files are duplicated and self-sufficient for each application's use, user access is fast and economical but cost of storage and maintenance is excessive. If all duplication is eliminated, either several files must be accessed to retrieve a full set of data required for a given application or files must be consolidated and will contain more data. In the latter case, sequential processing is less efficient, and random-access processing requires multiple accessing for selective data retrieval.

The problem is to determine the file organization that best balances the advantages and disadvantages. This problem is complicated by another database objective called *application independence.* If the number, nature, and frequency of applications were fixed, it would be possible to tailor file organization to optimize overall efficiency. But the goal is the opposite: to organize databases independently of applications and have them serve in common all applications and subsystems, both current and future. Application independence also implies that changes in application programs (like MRP) should not cause reorganization of the database, a major and costly undertaking.

How best to meet the requirements of efficiency in storage, file maintenance, and data retrieval is a technical system problem that, in practice, is solved by compromises between the conflicting objectives. Different database software packages embody different approaches to this solution.

When files become part of a system database, they serve as a common base for multiple applications and can no longer "belong" to individual departments in which data originate. File data are not "owned" by any one business function using the file contents. Data file management functions consist of:

1. File creation
2. File organization
3. File access
4. File updating (transaction processing)
5. File maintenance (changes and corrections)
6. Inquiry capability
7. Report generation

File Organization and Access

A file is made up of a set of records, each consisting of a number of fields containing pertinent data. Each record has a key, a field or data element, that serves as its identification by which the record is found in the file. An example is the part number in an inventory record file. Each file is organized in the sequence of its key numbers. File organization and file access problems arise when some data other than the key are to be retrieved. Such data, while they occupy the same field in each record, cannot be sequenced in their own right, and therefore it is not possible to access and retrieve them directly. They must be accessed indirectly, through one of the following:

1. A file scan (inefficient)
2. A separately maintained index
3. A linked record or chain

The problem and some solutions can best be illustrated using a telephone directory, a set of records of telephone users. Each of these records has three or four fields:

1. Name
2. Street address

3. Subdivision
4. Telephone number

The name is the key, and the file is organized in alphabetic sequence. Envision this file residing in computer random-access storage and pretend to be a computer program retrieving data. To learn the telephone number of a person named Morris, open the book somewhere beyond its middle. Since we are familiar with the key and alphabetic sequences, we know approximately where to look for the M's. Starting from the page we have open, conduct a file scan forward or backward until the desired record has been located.

The logic of this technique can be programmed for computers by setting up a subsidiary record or index (which we carry in our heads). This might indicate, for instance, the range of storage addresses (phone-book pages) for each letter of the alphabet. This index would be consulted first, and the subsequent scan would be accordingly limited. But the name, the key itself, is not a problem to find directly in computer systems; a number of techniques are available for minimizing scans or even accessing the record directly. The real problem arises when we seek other information such as:

1. Which names represent pharmacies?
2. Whose number is 322-7593?
3. What are all the names with a 48th Street address?

A second telephone directory, called "Yellow Pages," contains the answers to Question 1 arranged for easy access using another key—occupation or type of business. Computers make such record duplication unnecessary; in them, the yellow pages file would contain only the storage addresses of pharmacies but no names, street addresses, or telephone numbers, since these can be extracted from name-key records. Such record linkages are called *chaining*; the contents of one record are linked to the location of another. This will be described in more detail later.

If question 1 were frequent, an index to telephone numbers could be constructed by listing all pages on which the prefix 322 appears in separate records. This would create another small subsidiary file but would limit file search time. Problem No. 3 could be solved similarly. Obviously, more extensive indices limit the search time, but at the expense of storing a larger number of subsidiary files.

If a file is to serve multiple applications, including unspecified future ones, file organization and data retrieval problems are common and constant, involving a tradeoff between the economy of file storage and the efficiency of data retrieval.

File Maintenance

In the past, practically every manufacturing company worked with very poor-quality data; experienced people estimated that about half of all data used in planning and control had significant errors! Clerks manually maintained files of product design drawings and specifications, product structure (BoM), inventory status (on-hand and on-order), open order status for purchased and manufactured items, and processing information (routings, operation sheets, tooling). Unfamiliarity with the data, complex forms, little training, and many changes affecting the records ensured a high incidence of errors. Frequent physical inventories, poorly conducted, put even more errors into records than they removed.

Prior to MRP, manual inventory control systems depended on file data integrity; errors hampered their performance. However, these systems were inherently incapable of producing valid schedules and their users, therefore, improvised informal systems to make up for this deficiency. These hid the effects of file data errors in the formal system. Execution of plans depended on informal side records and expediting to get things made.

When powerful computers and MRP software programs were introduced into inventory management and production scheduling, it was recognized that these would be ineffective because of poor-quality data. It was clear that files must be overhauled, restructured, brought up-to-date, and maintained accurate. This file cleanup was part of every MRP implementation effort and, although massive, was usually effective in getting more accurate data.

Good file maintenance is extremely difficult; rarely are adequate resources allocated for this function. There are many changes, and some of these affect literally hundreds of records. Nonmanufacturing people don't realize how many components must be planned and controlled for even simple products. A wooden pencil, for example, has 10 component parts and materials:

Wood barrel (half)	Box (inner)
Graphite (lead)	Label
Ink	Ferrule
Eraser	Paint
Glue	Carton

Customer preferences require 5 hardnesses of graphite, 3 erasers, 2 types of barrels (round and hexagonal), 5 paint colors, 3 box sizes, 120 brand labels, and 3 shipping cartons. Each style of pencil needs a bill of material to specify which parts to use when making it. To produce a pencil takes 15 different operations, some manual and some machine, all of which need to be scheduled and monitored. Data on processing are even more difficult to maintain than bills of material. All of these files have to be maintained rigorously if MRP is to function properly.

All computer-based systems require effective file maintenance. Computer manufacturers and software suppliers have invested heavily in the development of file management programs (database software) to help users cope with the problem of maintaining file integrity. Their use requires adequate, competent staffing, often not budgeted initially and later viewed as an optional expense by managers under pressure to reduce costs. Errors resulting from economizing on file maintenance may not show up immediately but will eventually prove very costly in impaired system effectiveness.

The Complete Logical Record

An MRP program has a set of logically linked item inventory records, coupled with programs that accept data to maintain these records up-to-date. Each record consists of three portions or segments: item master data (header), inventory status, and subsidiary data. An example of a full item inventory record is shown in Fig. 7-1 and consists of

I. Item master data segment
   A. Item identity
   B. Item characteristics
   C. Planning factors
   D. Safety stocks
   E. Pointers to other files
II. Inventory status segment
   A. Gross requirements
      1. Control balance or past-due fields
      2. Time-phased data fields
      3. Totals
   B. Scheduled receipts
      1. Control balance or past-due fields
      2. Time-phased data fields
      3. Totals
   C. On hand
      1. Current on-hand fields
      2. Allocated on-hand fields
      3. Projected on-hand fields
      4. Totals (ending inventory or net requirements)

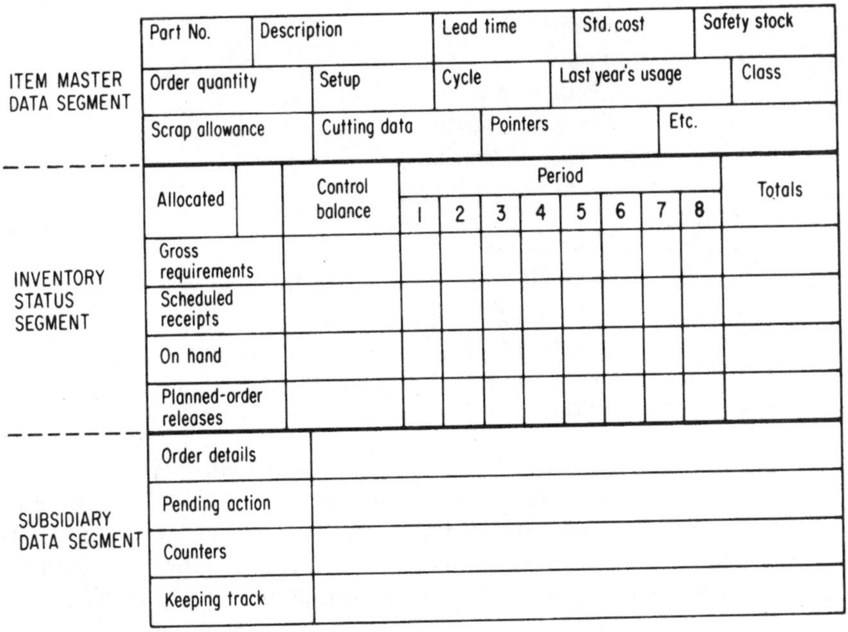
Part No.
Description
Lead time
Std. cost
Safety stock
ITEM MASTER DATA SEGMENT
Order quantity
Setup
Cycle
Last year's usage
Class
Scrap allowance
Cutting data
Pointers
Etc.
Allocated
Control balance
Period
1 2 3 4 5 6 7 8
Totals
INVENTORY STATUS SEGMENT
Gross requirements
Scheduled receipts
On hand
Planned-order releases
Order details
Pending action
SUBSIDIARY DATA SEGMENT
Counters
Keeping track

Figure 7-1. Logical record of an inventory item.

D. Planned-order releases
   1. Control balance or past-due fields
   2. Time-phased data fields
   3. Totals

III. Subsidiary data segment
   A. Order details
      1. External requirements
      2. Open (shop and purchase) orders
      3. Released portion of blanket orders
      4. Blanket order detail and history
      5. Other (user's choice)
   B. Records of pending action
      1. Purchase requisitions outstanding
      2. Purchase order changes requested (quantity, due date)
      3. Material requisitions outstanding
      4. Shop order changes requested (rescheduled due dates)
      5. Planned (shop) orders held up, material shortages
      6. Shipment of items requested (requisition, etc.)
      7. Other (user's choice)

C. Counters, accumulators
   1. Usages to date
   2. Scrap (or vendor rejects) to date
   3. Details of demand history
   4. Forecast errors, by period
   5. Other (user's choice)

D. Tracking records
   1. Firm planned orders
   2. Unused scrap allowances, by open shop order
   3. Engineering change actions taken
   4. Orders held up, pending engineering change
   5. Orders held up, raw material substitution
   6. Other interventions by inventory planner

These data collectively are termed the *logical records* (data logically related) as against the *physical records* (in computer storage) in possibly different formats and different locations. Data constituting logical records are not necessarily stored together; some may not be stored at all but recreated in the computer's memory when needed for computation and/or display.

Time-Phased Inventory Records

A separate time-phased record is established and maintained for every inventory item. Each record has the three segments listed in the tabulation above. The inventory status segment is either reconstructed periodically or kept up-to-date continuously, depending on whether regeneration or net change replanning (covered in Chapter 5) has been chosen.

The popular horizontal format for recording and displaying time-phased inventory status data, shown in Chapter 4 in several examples, consists of four rows of time-buckets:

1. Gross requirements
2. Scheduled receipts (open orders)
3. On-hand (current and projected by period)
4. Planned-order releases

This format accommodates all information essential for the proper handling of status data for MRP. It shows inventory status in summary form and contains added information implicit in the data displayed directly. This is the standard format favored by many MRP users and is the one commonly used in education and user instruction. It is said to "make the MRP logic obvious."

At users' option, however, the format can be expanded to provide more detail and/or state more information explicitly. Figure 7-2 shows both the compact and expanded formats for the same status data. The figure illustrates many possibilities of expansion; in practice, additions usually are limited to separate net requirements and/or planned-order receipt buckets.

A COMPACT FORMAT

Lead time: 3		Period 1	2	3	4
Gross requirements		10	15	75	17
Scheduled receipts		8		25	
On hand	72	70	55	5	-12
Planned-order releases		20			

B EXPANDED FORMAT

	Lead time: 3		Period 1	2	3	4
GROSS REQUIREMENTS	From parent items		10	10	15	12
	Service part orders			5		5
	Interplant				60	
	Total		10	15	75	17
SCHEDULED RECEIPTS	Supply source A				25	
	Supply source B		8			
	Total		8		25	
PLANNED-ORDER RECEIPTS						20
NET REQUIREMENTS						12
ON HAND	Stockroom #1	50	40	30		8
	Stockroom #2	22	30	25	5	
	Total	72	70	55	5	8
PLANNED-ORDER RELEASES			20			

Figure 7-2. Time-phased record: compact and expanded formats.

Optional Fields

The inventory status segment of the record may include a field labeled "Allocated on hand," in addition to those shown in Fig. 7-2. Allocations, discussed in Chapter 5, indicate the quantity of an item earmarked for a parent order released for production but for which the material requisition has not yet been filled; allocated parts "belong" to the respective parent orders but are still physically on hand in the stockroom because of the time gap between order release by planners and issuing requisitioned material from the stockroom.

A single allocation field in inventory records implies that parent orders normally will be released during the planned period and not before or after, and that all component items are required at the time the parent order is released. Both conditions are common, and the single allocation field is standard. In exceptional cases, when orders sometimes may be released prematurely or when requirements for components of an assembly are staggered over long assembly lead times, allocated quantities are time-phased; an example of this appears in Fig. 7-3. Note that use of a single allocation field (showing $10 + 15 = 25$) would distort the component item's projected on-hand schedule; Figure 7-3*B* would then show 15 in period 1 and 3 in period 2.

Another optional field, "Past due" in Fig. 7-4, is a time-bucket immediately preceding the first period commonly provided for this purpose in each row. It is convenient to use this for current on-hand data also; here "past due" has no meaning. Past-due quantities are recorded where appropriate. In the computation of net requirements, past-due gross requirements and receipts must be considered. In Fig. 7-4*A*, an assumption is made that the late order for 10 will catch up in the first period. Alternatively, past-due quantities can be included in the first-period bucket, as shown in Fig. 7-4*B*, but this violates a cardinal rule of MRP, "Make the logic obvious." Either way, the net requirements computation must produce identical results.

If master production schedules driving MRP contain past-due buckets, very many component records will show past due. It is unlikely that even the most energetic corrective actions will overcome all of these. Successful companies using MRP make strenuous efforts to avoid past-due data in the MPS.

Past-due data occur when execution fails to meet planned performance. This is expected and normal; perfect plans are never made, and execution is never completely successful. The ability of MRP to reveal the needed corrective actions is its greatest power. Recording a planned order release as past due, however, when there is insufficient lead time to get back on schedule, as shown in Fig. 7-5, makes little sense. The only sensible action

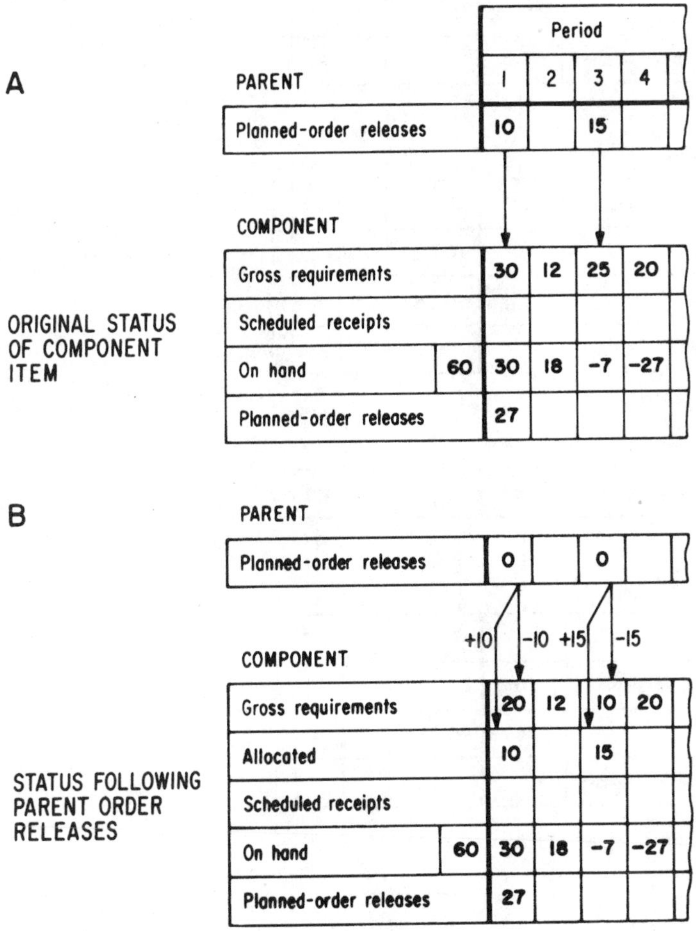

Figure 7-3. Time-phasing of allocated quantities.

is to correct for delinquent planner performance by showing order release in the current period. Chapter 11 contains a discussion of corrective actions in this situation.

It is possible to time-phase past-due quantities and to display them in the inventory record, as shown in Fig. 7-6. A good argument can be made that this shows *relative priorities* among past-due orders competing for scarce materials and capacity. Several companies have used this successfully in times of heavy overload. This capability, of course, is no substitute for having adequate resources to handle all orders. It just reduces, never eliminates, the pain and expense of late orders.

A

PARENT	Past due	1	2	3	4	
Planned-order releases	12	5		18	20	

COMPONENT

	Past due	1	2	3	4	
Gross requirements	12	5		18	20	
Scheduled receipts	10	15				
On hand	36	44	44	26	6	
Planned-order releases						

B

PARENT						
Planned-order releases	12	17		18	20	

COMPONENT

Gross requirements	12	17		18	20	
Scheduled receipts	10	25				
On hand	36	44	44	26	6	
Planned-order releases						

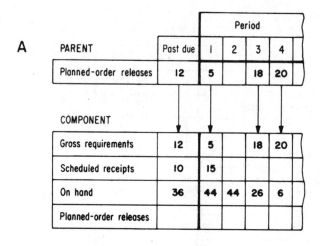

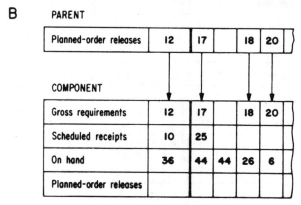

Figure 7-4. Treatment of quantities past due.

		Period				
Lead time: 5	Past due	1	2	3	4	
Net requirements					25	
Planned-order releases	25					

Figure 7-5. Past-due order release caused by insufficient lead time.

	Weeks past due				Week				
	4+	3	2	1	34	35	36	37	
Scheduled receipts	6			50			60		

Figure 7-6. Time-phased quantities past due.

In net change MRP programs, the function of the "past-due" field is taken over by the control-balance bucket. As is shown in Chapter 5, both delinquent and premature performance can be highlighted.

Another optional field in the status segment of the inventory record is a "total" bucket which appears at the end of each row. It is used for reconciliation purposes or validity checks; the computer can be programmed to carry out these checks whenever net requirements are recomputed. For example:

Current on hand:		115
Total scheduled receipts:		100
		215
Allocated on hand:	35	
Total gross requirements:	380	–415
Total net requirements:		–200
Total planned orders:		225
Planned coverage excess:		25

This type of reconciliation will detect errors in status data or highlight unusual conditions calling for possible action. For example, if the current on-hand and scheduled receipts total is significantly higher than the total of allocated and gross requirements, coverage is excessive and should be investigated. Possible causes are a reduction in gross requirements from customer order cancelations or an engineering change, an overrun of a work order, or an error in reporting receipts of another item.

Updating Inventory Records

Inventory status data are maintained by processing (posting) *inventory transactions* to item inventory records. An inventory transaction is any event that changes inventory status. *External* inventory transactions are reported to the system; *internal* transactions are generated by the system itself in running MRP. Reports of events that do not affect inventory status but are posted to the subsidiary data segment of the record are called *pseudotransactions*.

The entries that the program processes to update inventory records are

1. Inventory transactions
2. User-controlled exceptions to regular processing logic
3. Pseudotransactions
4. Final assembly schedule entries

5. Error-correction entries
6. File-maintenance entries

Inventory transactions modify the inventory status of items. A transaction may also cause subsidiary records to be processed in net change MRP programs if it requires updating of component-item status, thus affecting multiple inventory records. A transaction may report a normal or planned event, such as a stock receipt, or an unexpected event, such as a stock return. While both may have the same physical effect, the regular processing logic must be modified (as will be shown later) to register unexpected events.

User-controlled exceptions to regular processing logic provide means for planners to override the normal MRP logic when human judgment is required to evaluate and solve problems. There are several types of commands that MRP can be programmed to obey; one is a "hold" command to prevent a (mature) planned order from being issued, perhaps because of a contemplated substitution in raw material; a second is a "scrap" command to MRP to avoid release of a new order if its quantity is smaller than the scrap allowance of an existing open order; a third is a "firm planned order" command, which freezes a planned order in place. The use of this last command is discussed in Chapter 3.

Pseudotransactions do not affect inventory status; they are entries to the subsidiary data segment of the item inventory record. Examples are a purchase requisition issue (status will be affected only upon the release of the purchase order), a change in an open order number, and recording a subcontractor's work order (receipt of the material will add to the item's status).

Final assembly schedule entries are needed when end products are assembled to order or have several customer options and do not appear in the master production schedule. The MPS for such products is best stated in "standard modules," explained later in this chapter, rather than end-item models. The final assembly schedule states specific product models; the components required to support it may be allocated in their inventory records. Some companies use such a program to enter customer orders and define major component requirements. In companies practicing Just-in-Time, a day's or shift's end-item production is translated into high-level components consumed, and this total is entered into each component's inventory records in lieu of individual stock disbursement transactions. Called *backflushing,* this should be done only when processing lead times are short, bills of material highly accurate, and tight discipline exists in reporting end-item production promptly and accurately.

Error-correction entries may or may not affect real status. They are distinguished by special transaction codes from other transactions that may have the same effects. For example, an inventory planner releases an order for item A but erroneously reports it to MRP as item B, thus omitting a released order from A's data and adding a fictitious open order to B's data. The error to B is corrected with an entry that reverses the previous transaction, rather than by an order-cancellation transaction. The effect on B's data is the same, but the distinction is made to keep proper records of causes of transactions. The release of the order for A must be made correctly also, of course.

File maintenance entries update the item master data segment (header) of the item inventory record for changes in attributes such as description, standard cost, or ABC classification, or for changes in planning factors such as lead time or scrap allowance. File maintenance entries do not affect inventory status and do not trigger the standard MRP replanning process.

Transaction Types and Effects

Designers of MRP inventory control programs must decide how many different types of transactions are to be recognized, how they are to be coded, and how they will be processed. Choices are virtually unlimited; dozens of transaction types are recognized in MRP programs. While there is almost no limit to the number of different transaction types that may be used, these can have only a limited number of effects on inventory status.

Several different transaction types will affect inventory status the same way; for example, a stock receipt of an overrun on a production order, a customer return, and an upward inventory adjustment resulting from a physical count will all increase the quantity on hand and reduce net requirements. Each of the ten transactions in the following list has one, and only one, effect on item inventory status, as seen here.

External Transactions Affecting One Record

1. Change quantity of gross requirements
   Secondary effect: recompute projected on-hand and planned-order releases
2. Change quantity of scheduled receipt
   Secondary effect: recompute projected on-hand and planned-order releases
3. Reduce scheduled receipt and increase quantity on hand
4. Change quantity on hand
   Secondary effect: recompute projected on-hand and planned-order releases

5. Reduce quantity on hand and reduce gross requirements
6. Reduce quantity on hand and reduce quantity allocated

External Transactions Affecting Multiple Records

7. Change quantity of planned-order release (parent record) and change quantity of gross requirements (component records) *Secondary effect*: recompute projected on-hand and planned-order release in component records
8. Reduce quantity of planned-order release and increase scheduled receipts (parent records); reduce gross requirements and increase quantity allocated (component records)
9. Increase quantity of planned-order release and reduce scheduled receipts (parent record); increase gross requirements and reduce quantity allocated (component records)

Internal Transactions Affecting Multiple Records

10. Change quantity of planned-order release (parent record) and change quantity of gross requirements (component records). *Secondary effect*: recompute projected on-hand and planned-order releases in component records

Effect No. 1 (change quantity of gross requirements) either increases or reduces the contents of one or more gross requirements buckets. Effect No. 1 results from transactions reporting increases, reductions, or cancellation of demand for an item originating from external sources. Examples are orders or forecasts for service parts and interplant transfers of components. A change in the timing of a gross requirement is effected by reducing the quantity in the original bucket and increasing the quantity in the new bucket.

Effect No. 2 (change quantity of scheduled receipt) results from increasing, reducing, canceling or rescheduling an open order. A purchase-order increase, a supplier's overshipment, and a change in the order due date will have this effect. Rescheduling, as with gross requirements changes, involves reducing the contents of one bucket and increasing that of another.

Effect No. 3 (reduce scheduled receipt and increase quantity on hand) is caused by a full or partial stock receipt of material ordered. This does not apply to an unplanned receipt for which no order had been placed previously, nor to the quantity of an overrun or overdelivery. Unless delivery is premature, neither the projected on-hand nor the planned-order release schedules need be recomputed.

Effect No. 4 (change quantity on hand) is the result of transactions that increase or reduce the quantity on hand without affecting any open orders. Stock returns, overdeliveries, inventory adjustments, and unplanned disbursements have this effect. Such unanticipated changes in the quantity on hand cause recomputation of the projected on-hand schedule and, consequently, the planned-order release schedule.

Effect No. 5 (reduce quantity on hand and reduce gross requirements) results from a disbursement or shipment of an external service part or interplant order. These have no secondary effects on the other status data in the record.

Effect No. 6 (reduce quantity on hand and reduce quantity allocated) is caused by a planned disbursement of a component against a parent order. As stockrooms fill material requisitions or picking lists, the reporting transactions reduce the item quantities on hand and allocated.

Effect No. 7 (change quantity of planned-order release in the parent record and change quantity of gross requirements in the component records) is a result of intervention by the inventory planner, solving problems by changing the quantity or timing of a planned order. The planner avoids recurrence of this problem by "freezing" this change so that the MRP system will not recompute or reposition this planned order the next time net requirements change. The transaction reporting this intervention to the system is the firm planned order that affects more than one record. The change in the planned-order schedule affects the gross requirements of all components of manufactured items and causes their status to be recomputed. Since purchased items have no components in the user's MRP program, such transactions never affect other inventory records. An exception to this is the case where the customer supplies material to the supplier.

Effect No. 8 (reduce quantity of planned-order release and increase scheduled receipts in parent records; reduce gross requirements and increase quantity allocated in component records) is initiated by release of a planned order, which the transaction converts to an open order (scheduled receipt) in the parent item's inventory record. This transaction also affects component records, where gross requirements are reduced and allocated quantities increased.

Effect No. 9 (increase quantity of planned-order release and reduce scheduled receipts in the parent record; increase gross requirements and reduce quantity allocated in component records) nullifies a previous order-release transaction. This happens when inventory planners decide to cancel or change a previous release of an order. Once work on a shop order has started, this is rare.

Effect No. 10 (change quantity of planned-order release in the parent record and change quantity of gross requirements in component records) is the only internal transaction affecting multiple records. It is caused by MRP replanning, which changes a parent planned-order schedule, requiring different gross requirements of component items. This effect is identical to No. 7, except that here the "transaction" is generated by the system internally in the requirements planning explosion.

Different transaction codes may be used for several entries that are logically identical, having the same effects on inventory status. The reasons for creating a set of transactions larger than the essential minimum are the desirability of being able to log transaction histories (and create an audit trail by recording their sources and reasons) and the ability to provide different treatments of these various transactions in the subsidiary data segment of the inventory record.

MRP program designers and users are free to create, without limits in number and type, transaction sets best suited to their needs. All possible sets cannot be covered here, but Fig. 7-7 lists the basic transactions that must be reported to the system from external sources.

Bills of Material

Unlike order point, MRP uses bills of material (BoM) as its basis; these define the relationships of assemblies (parent items) and their components. BoM acquired a new function with MRP, *becoming a framework on which the entire planning process depends.* As a vital part of MRP, BoM must be accurate and up-to-date if its outputs are to be valid. Parent and component items omitted from BoM will not be available when needed, and those in BoM not belonging there will add to excess inventory.

In addition, BoM must be unambiguous and structured properly for planning. The term *bill of material structure* refers to the arrangement of parents and components within the BoM file, not to the organization of this file in some storage medium. Bill-of-material processor programs, software packages mentioned in Chapter 3, do not structure them; they edit, organize, load, maintain, and retrieve BoM records already properly structured by knowledgeable people to serve MRP's needs.

In the past, bills of material were provided by design engineering to describe products (and the way they were put together) to others in the organization needing such information. These included production, material planning, purchasing, quality control, and cost accounting. Each requested, but did not always receive, minor modifications for its special uses.

Transaction type	Sample code
Requirements from external sources	
▪ Enter	A1
▪ Cancel	A2
▪ Increase	A3
▪ Reduce	A4
▪ Change timing	A5
Order release	
▪ Place order	B1
Change in scheduled receipts	
▪ Cancel order	C1
▪ Increase quantity	C2
▪ Reduce quantity	C3
▪ Reschedule	C4
▪ Scrap in process	C5
Receipts	
▪ Scheduled receipt	D1
▪ Unplanned receipt	D2
▪ Inventory adjustment upward	D3
▪ Stock return	D4
Rejected shipments	
▪ Reject	E1
▪ Cancel rejection	E2
Disbursements	
▪ Scheduled disbursement	F1
▪ Unplanned disbursement	F2
▪ Inventory adjustment downward	F3
▪ Scrap in storage	F4

Figure 7-7. Basic transactions from external sources.

The following checklist includes the major uses of BoM:

1. Improving forecasting of optional product features.
2. Making order entry simple and easy, expressing customer orders either in model numbers or as a configuration of optional features.
3. Stating the master production schedule in the fewest possible number of items. These items will be products, major subassemblies, or standard end-item modules.
4. Planning subassembly priorities so orders for subassemblies are released at the right time, with valid due dates kept up-to-date.
5. Assembly scheduling, including all components required to build parent items.

6. Specifying how products are built, including added subassemblies and semifinished items for economical production and for replacement parts.
7. Product costing, which can ignore minor differences not affecting costs significantly.
8. Efficient computer file storage and file maintenance.

Bill-of-Material Structuring

BoM issued and maintained by engineering departments rarely meet the needs of all users; they have to be restructured without affecting the integrity of product specifications and in such a way as to permit modifications complying with engineering changes. The severity of the restructuring problem varies from company to company, depending on the complexity of its products and the nature of its business. BoM structuring must include

1. Assignment of item identities to
   a. Eliminate ambiguity
   b. Define levels of manufacture
   c. Handle transient subassemblies
2. Product model designations
3. Modular bills of material
   a. Disentangling product option combinations
   b. Segregating common from unique parts
4. Pseudobills of material
5. Interface with order entry

MRP requires BoMs in which *each item to be planned must be uniquely identified.* One part number must not identify two or more different items. Product designers, industrial engineers, cost accountants, inventory planners, and production workers might each prefer to assign them differently; part numbers have many uses including identifying products to customers, raw materials to suppliers, service parts to field service people, and many more. Our concern is with MRP needs.

MRP requires unique identities for every item it plans, for each item obtained by purchase orders on suppliers, for each made on work orders in a manufacturing facility, and for every item to be carried in inventory. Design engineers are concerned primarily with functions of components and products in assigning part numbers; they view the bills of material as the end of their design task. Others in manufacturing view them as the beginning of their work, and often need additional part numbers and different structuring. Examples are added part numbers for service parts made differently from components used in production, added subassemblies that make processing easier and less expensive, and modules that improve planning and control.

Identifying numbers must *define the components of a parent unambiguously*; the same subassembly number must not be used for two or more different sets of components. This sometimes happens when a product is redesigned; instead of creating a new bill of material for the revised assembly, the original parent BoM number is retained if engineers believe that form, fit, and function are the same. Others may need different numbers for their uses.

Computer illiterates might think that the tremendous speed of a computer enables it to find anything in a file almost instantaneously and that considerations of file organization, maintenance, and data retrieval are not very important. They *are* important, however, because even microseconds add up to significant amounts of time with the masses of data encountered in even small manufacturing companies. If a telephone directory containing one million names (many American cities have listings that large) were to be scanned from a random-access storage device at 20 milliseconds per access to an individual record, the total search time would amount to five hours!

Stages of Manufacture

Bills of material should reflect, through their level structure, material movements into and out of stock, not necessarily into a stockroom but into a state of completion. When a fabricated part or a subassembly is completed, it is considered to be "on hand" until withdrawn as a component of an order for a parent. MRP assumes, as discussed in Chapter 3, that each inventory item under its control goes into and out of stock, and also assumes that the BoM accurately reflects this.

Planning BoMs must specify not only the composition of a parent but also the process stages in its manufacture. BoM product structure levels must define each step in the buildup of the product so that MRP can establish, using planned lead times, the timing of order releases and completion.

Some MRP users question assigning separate identities to semifinished and finished items, where the finishing operations are minor. A die casting that is first machined and then receives one of three finishes, painted, bronze, or chrome, is shown in Fig. 7-8. The three finished items (Parts 3, 4, and 5) must have separate identities *if their order priorities are to be planned by MRP and if they are to be identified in stock.*

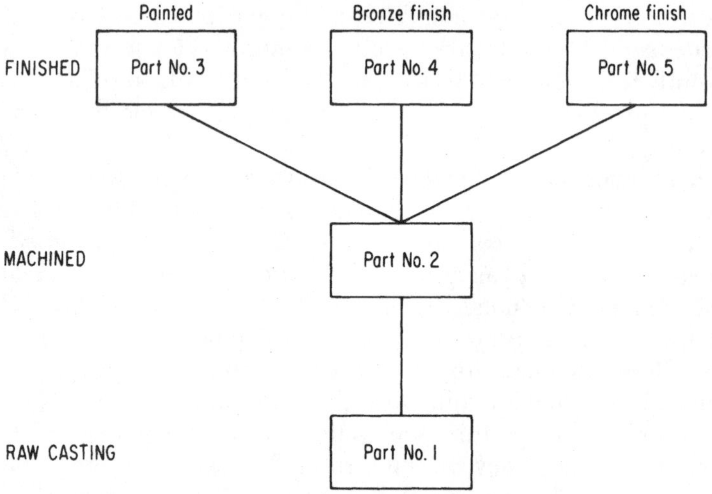

Figure 7-8. Unique identity of semifinished and finished items.

Transient Subassemblies

An item-identity problem almost the opposite of the preceding one is the transient subassembly, sometimes called a *phantom item.* Such assemblies never see a stockroom; they are immediately consumed in the assembly of parent items. An example of this is a subassembly built on a feeder line that flows directly into the main assembly line. If the subassembly carries a separate identity in the BoM, MRP will treat it the same as any other subassembly moving into and out of stock. This requires all receipts and disbursements to be reported using artificial transactions to update time-phased inventory records. Such transactions, unrelated to actual movement of materials, are difficult to record accurately using conventional means.

Transient subassemblies would not have to be identified in BoMs at all if there were never an overrun, a customer return, or service-part demand. If these happen, they must be identified separately in BoMs and their inventory status maintained. In net change MRP this poses a problem, because all transactions for transient subassemblies would have to be reported continuously to maintain inventory records up-to-date. This serves no purpose in order releases, order completions (receipts), and issues.

Fortunately, a technique called the *phantom bill* makes this unnecessary; transactions related to work-in-process orders do not have to be reported and posted, although those for stock room receipts and issues do. Using this technique, MRP picks up and uses any transient subassemblies on hand; when none are on hand, MRP bypasses the phantom item's record and goes from its parent to its components directly. Service-part requirements also can be entered into the record and will be correctly handled.

This example shows how the technique works. Assembly *A* has a transient subassembly *B* as one of its components, and part *C* is a component of *B*. Item *B*, the phantom, is sandwiched between *A*, its parent, and *C*, its component, and is treated as follows:

1. Lead time is specified as zero.
2. Lot sizing is Lot-for-Lot.
3. The BoM or the item record is coded to identify *B* as a phantom and to ensure special treatment.

The difference between the special treatment of phantoms and regular MRP update logic can be seen in Figs. 7-9, 7-10, 7-11, and 7-12. Figure 7-

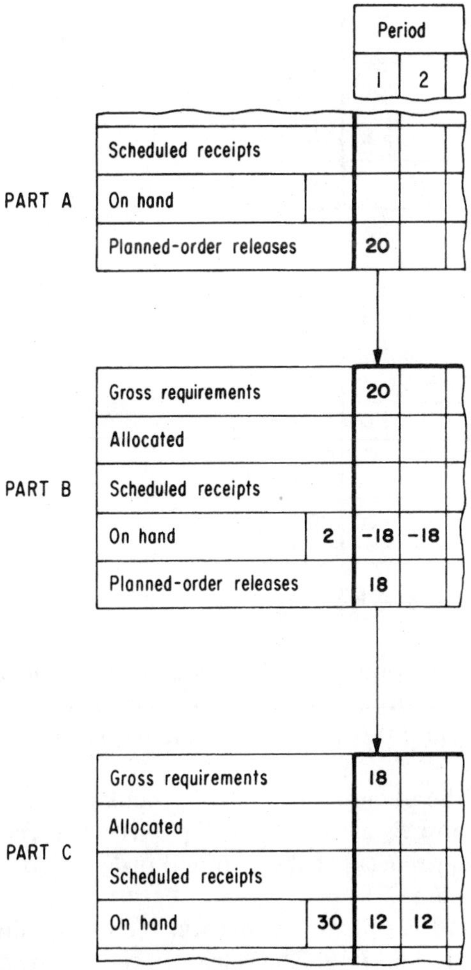

Figure 7-9. Transient subassembly *B*, its parent and component.

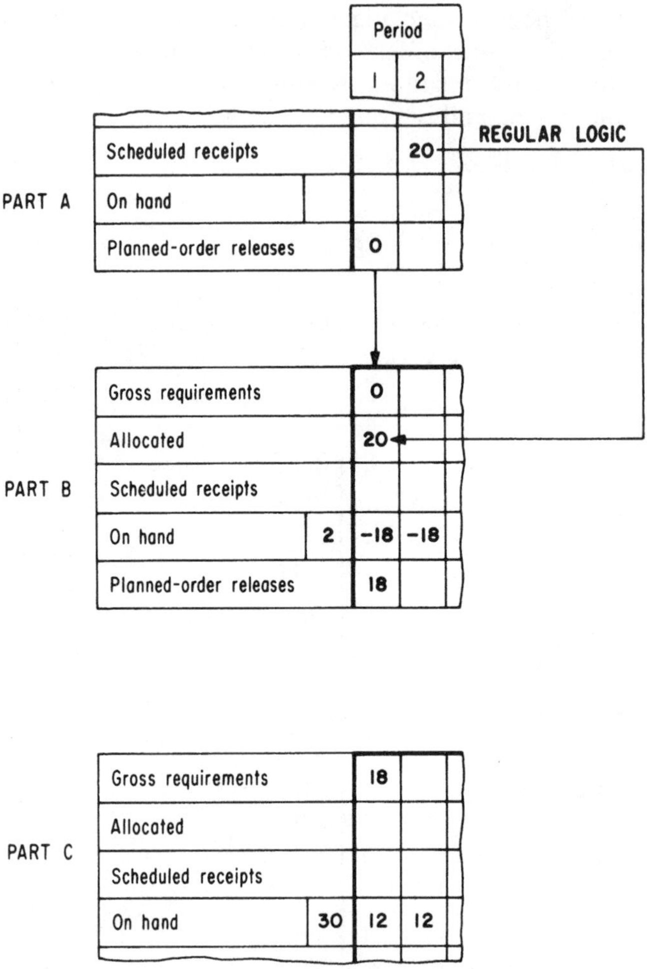

Figure 7-10. Regular update logic, following release of order for item A.

9 shows inventory status of items *A, B,* and *C.* The zero-lead-time offset on *B* places the planned-order release for 18 pieces in the same period as the net requirement, and this in turn puts the requirement for 18 *C*'s in the same period.

In the absence of the phantom code for *B,* regular MRP logic updates records of items *A* and *B,* as shown in Fig. 7-10. The record for *C* is unchanged. Following release of the planned order for *B, C*'s record is updated as shown in Fig. 7-11.

If item *B* were coded as a phantom, all three records would be updated in one step, depicted in Fig. 7-12. The release of the planned order for *A,*

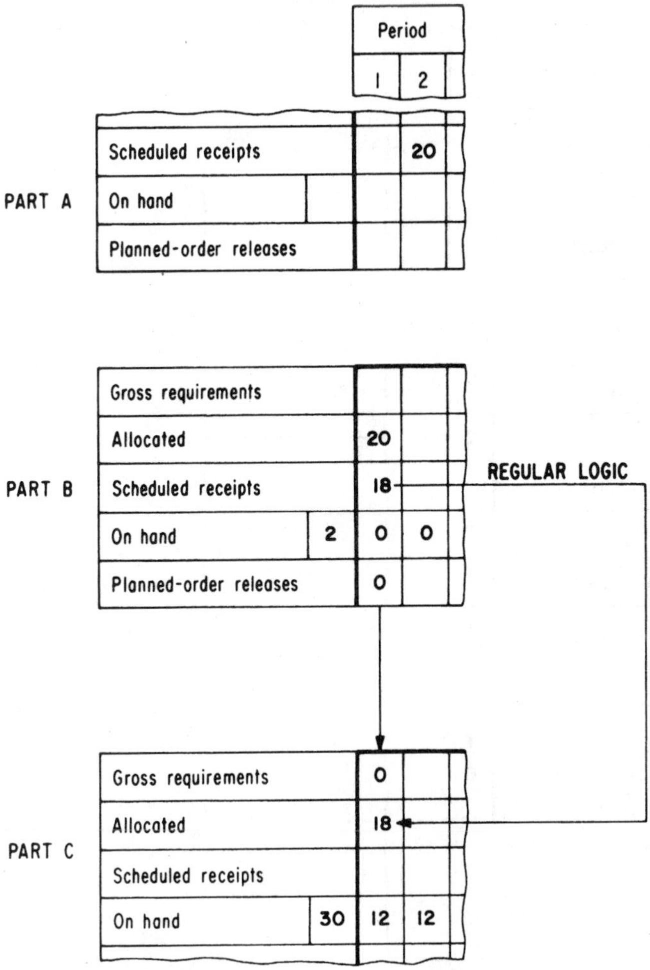

Figure 7-11. Regular update logic, following release of order for item B.

which normally would reduce only the gross requirements for *B* as in Fig. 7-10, now also reduces gross requirements for *C* (as though *C* were a direct component of *A*). The two units of *B* in stock (perhaps a return from a previous overrun) are applied by allocation to help meet the gross requirements for *A*; the remainder of 18 is allocated to *C*. Study of these examples shows that the phantom logic is nothing more than a different treatment of allocation combined with zero lead time and Lot-for-Lot ordering.

The phantom bill technique is most useful with net change MRP programs. In regenerative MRP, not posting transactions to phantom

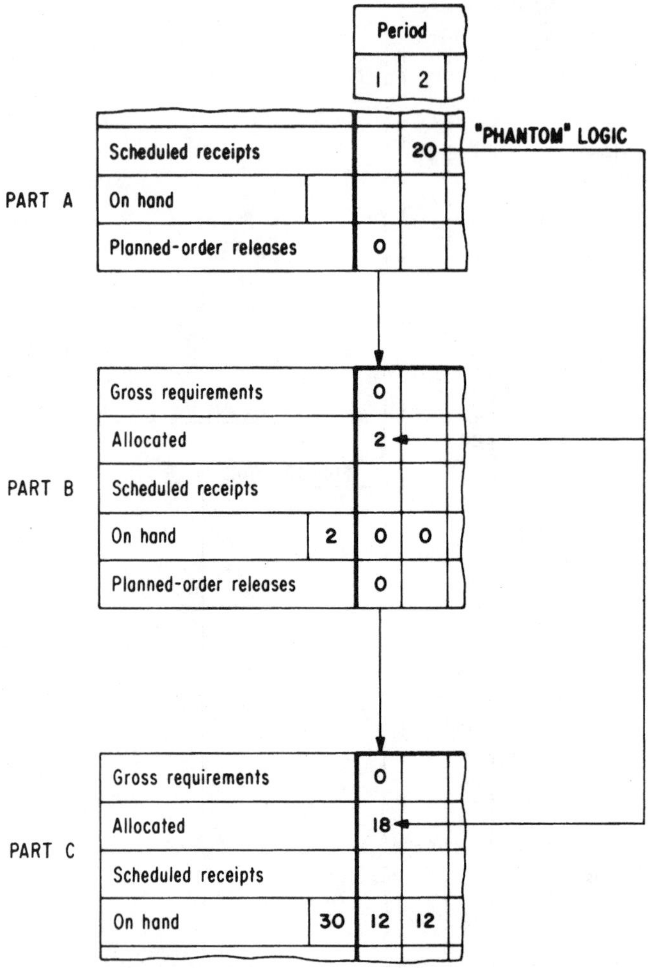

Figure 7-12. Simultaneous updating of item B's parent and component records.

records to cover assembly activities may be worthwhile but is not crucial, because a planned-order release does not update component gross requirements. Hence, the problem of rebalancing (realigning) the planned-order and gross requirements data of the three records does not arise.

Following the planned-order release of the transient subassembly's parent, the next requirements planning run will wipe out both the gross requirement and the planned-order release for the transient subassembly. Coding inventory records of transient subassemblies can suppress notices for planned-order releases or flag them to be disregarded.

The problem then becomes one of requisitioning components for Part

A's order. It is solved by modifying the procedure that generates material requisitions. When some transient subassemblies are on hand, two requisitions are generated, one for the quantity of the transient subassembly on hand and one for the balance needed from its component items. In Fig. 7-12, these quantities are 2 and 18, respectively.

Product-Model Designations

Product lines usually consist of a number of models, often in families of similar products. Marketing or sales people normally think of products and forecast sales in specific models; production thinks of models; the final assembly schedule is for models, and the master production schedule may also show models. Model numbers may be unique identifiers and yet not be adequate for material requirements planning, because they fail to provide complete product definition.

Products having many optional features for customers to select provide a large number of possible option combinations in specific models; many of these may never be ordered by a customer. For example, a farm tractor family might have the optional features shown in the table on page 176.

With these options, it is possible to build

$$3 \times 3 \times 2 \times 2 \times 2 \times 2 \times 2 \times 2 \times 3 \times 2 \times 2 = 6912$$

tractors with no two identical; each would have a unique configuration of optional features. A complete set of model designations would contain 6912 different numbers. If a single digit or letter were used for each of the characteristics, model designations would contain 11 characters. Sales, marketing, advertising, and many other people prefer shorter numbers; obviously, shorter significant-digit numbers will not provide complete information on model options and will be useless for MRP.

Specific details of options desired by customers must be provided for customer-order entry, pricing, and promising deliveries. These details will define the specific modules required to build the products customers want; this information can be used to test how well earlier planning and execution are providing components (and other resources) using allocation techniques described in Chapter 9. These details are needed also for final assembly scheduling, production, and shipping, all part of the execution phase of manufacturing.

In the planning phase, however, it should be obvious that attempts to forecast demand for a multitude of specific combinations of many options are doomed to fail. Developing master production schedules and setting up bills of material for many models customers may never order is a gross

Function	Options
Wheel arrangement	Four-wheel construction
	Three-wheel construction, single front wheel
	Three-wheel construction, double front wheel
Fuel	Gasoline
	Diesel
	LP gas
Horsepower	56 hp
	68 hp
Transmission	Stick shift
	Automatic
Steering	Mechanical
	Power
Rear platform	Regular
	Low
Axles	Standard
	High-clearance
Hitch	Mechanical
	Hydraulic
Power takeoff	With type A
	With type B
	Without
Radiator shutters	With
	Without
Operator cab	With
	Without

waste of effort. Reducing the number of options or limiting combinations of options would improve the quality of forecasting and simplify master scheduling, but might alienate customers. There is a better solution: *apply MRP to plan for major subassemblies or sets of parts, commonly called "modules."* This approach results in great improvements in forecasting, master scheduling, and planning for the procurement, fabrication, and subassembly of components and provision of other resources.

Modular Bills of Material

Modular BoM are used to simplify and improve master scheduling. Instead of product models, they describe sets of components for product functions, options, attachments, or common use in many products. Modularizing rearranges end-item bills of material, grouping components into useful sets called modules, with two different objectives:

1. Disentangling combinations of optional product features
2. Segregating components common to many parents from those unique to one parent or peculiar to some function

The tractor example in the previous section helps to illustrate how these two objectives are met. It is possible to set up and store a bill of material for each of the 6912 different tractors, but it is neither practical nor necessary. Many of the possible configurations may never be sold during the life of the product; their BoM would never be used. Furthermore, a few design improvements and engineering changes could add myriad additional bills. For example, if design engineers at marketing's request add an option of special fenders with mudguards, the number of possible option combinations doubles from 6912 to 13,824.

Developing valid master production schedules for so many models is impossible. If the manufacturer produces 300 of this type of tractor per month, how many of which of the 6912 (or 13,824) models should make up the 300? Higher volumes do not eliminate the problem. Even planning for 10,000 per month of a product family with only 100 possible option combinations is impractical, and a better solution is available.

This solution is to forget, in the planning phase, end products having combinations of options and attachments; forecast and plan each of their top-level components separately. In the example, if the production rate is 300 tractors per month, each month schedule 300 sets of components common to all models (fenders, hoods, rear wheels, etc.) in a BoM for this module.

There are two transmission options. Past demand has averaged 75 percent stick shift and 25 percent automatic, and a monthly schedule of 225 and 75 units, respectively, is reasonable. These may be good averages, but customer orders in any one month are unlikely to break down exactly that way; some safety stock would be desirable. As mentioned in Chapter 6, the best way to plan safety stock *if desirable in MRP* is at the master production schedule level. Each type of transmission could be "overplanned" based on past variations in actual orders, adding say 25 to each MPS only in the first period to avoid distorting capacity requirements. This approach could be followed for other optional features.

For the tractor functions listed above, this would require setting up only 25 BoMs and developing 25 MPS as follows:

Basic tractor:	1
Wheel arrangement:	3
Fuel and horsepower:	6
Transmission:	2
Steering:	2
Rear platform:	2
Axles:	2
Hitch:	2
Power takeoff:	3
Radiator shutters:	1
Operator cab:	1
Total	25

This total compares with 6912 if each tractor configuration had its own BoM. If the engineers add special fenders, instead of doubling the number of bills, it would add only one BoM to the file (regular fenders already have one). Both regular and special fenders need to have MPS. Before restructuring BoMs for modularity, manufacturers usually have complete BoMs for a few of the possible configurations covering models already built; these may be already or could be adapted for other models by adding and subtracting components to account for configuration differences. This is satisfactory for many uses of BoMs but not for MRP.

Percentage bills of material have been suggested for use with options to minimize the number of MPS to manage. In the transmission example earlier, the BoM for the parent tractor would show both transmission options with a quantity per assembly of 0.75 for the stick shift and 0.25 for the automatic. Similarly, for large families of products, a family BoM would include each model with the percentage it has of total family requirements, possibly increased for safety stocks. This has good applications where the costs of planned items are low, for example, package options for bulk products like aspirin or cold remedies sold in many different package sizes.

In most other applications, the problems with percentage BoM outweigh the advantages. It is difficult to make valid customer delivery promises based on MPS and percentage BoM. Changes in overplanned amounts require changing BoM structure data. Many software programs cannot handle fractional requirements resulting from low usages. The technique fails when the mix of options or models is not constant over the planning horizon, when some items have very small percentages, or when it is desirable to alter mixes for seasonal products.

The Modularization Technique

The technique of restructuring end-product BoM into a modular format is illustrated using the tractor example, simplified to only two optional features; customers can choose only stick-shift or automatic transmission, and mechanical or power steering. Figure 7-13 shows four BoM; the first combines stick shift and mechanical steering, the second, stick shift with power steering, and so on. In the figure, end-product model numbers (12-4010, 12-4020, 12-4030, and 12-4040) are on level 0 of their respective BoM. Level-1 components may also be assemblies, but their components are omitted for simplicity.

Restructuring these bills of material involves grouping level-1 components into logical modules. For example, A13 is common to all models and would be assigned to the "common" group. C41 is found in stick-shift/mechanical and stick-shift/power models, not with automatic transmission models, indicating that it should be associated with the stick-shift option. L40 is used only with mechanical steering. The remaining Level-1 component items are similarly assigned to groups, as shown in Fig. 7-14.

The level-1 components, D14, H23, J32, and N44, remain unassigned; each is used with only one of the option combinations. For these the process is carried one step further, as shown in Fig. 7-15, looking at their

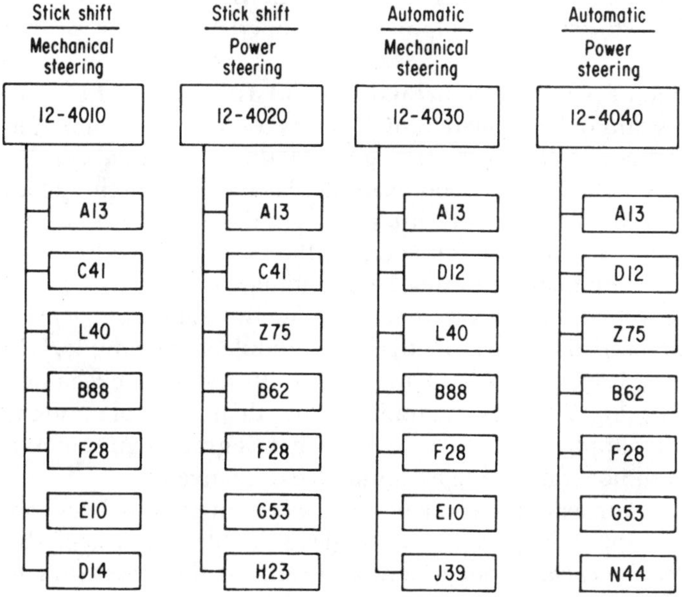

Figure 7-13. Bills of material for four tractor models.

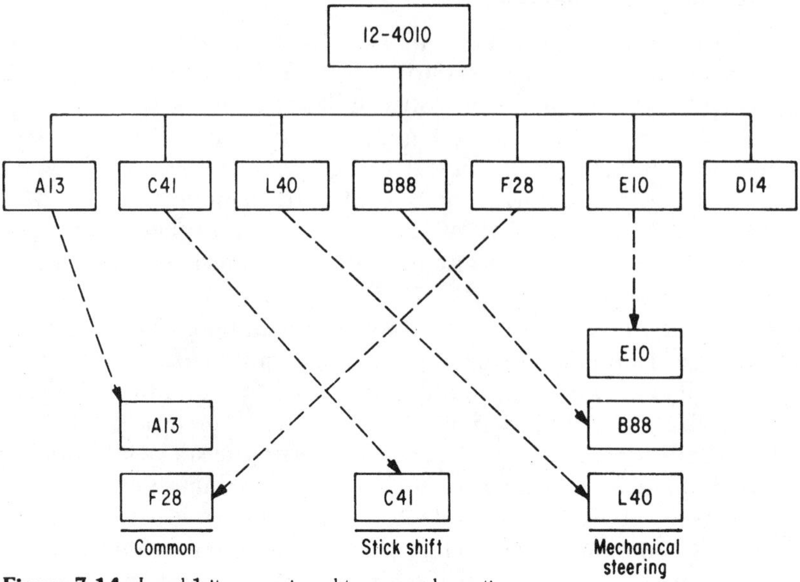

Figure 7-14. Level-1 items assigned to groups by options.

Level-2 components and assigning them to options. The final result is shown in Fig. 7-16, with all components grouped under the proper options.

The technique of parent BoM breakdown and assignment of components to options cannot be applied to single parts used with combinations of options. The best solution is redesign, but this will not always be feasible. The engine block is a good example. Heavy-duty engine-block castings are used for diesel power, different from those for gasoline and LP gas, and the horsepower option entails different cylinder bores; fuel and horsepower combinations cannot be disentangled.

Expensive combinations must have separate modules, but inexpensive components can be assigned to more than one module. For example, D14 in Fig. 7-13 could appear in both the stick-shift and the mechanical steering modules (Fig. 7-16), overplanning rather than underplanning it. Since MRP always applies on-hand inventory against gross requirements, overplanning will not continue to accumulate excess inventory.

The tractor in our example can have four-wheel or three-wheel construction, and an option in the latter is a single or double front wheel; this is *an option within an option* (a suboption) and calls for the establishment of three modules:

1. Items common to all three-wheel options

2. Items unique to the single-wheel suboption
3. Items unique to the double-wheel suboption

As mentioned above, forecast errors on options make it reasonable to use safety stock, and these are frequently overplanned. Overplanning suboptions must be done carefully to avoid pyramiding inventory. For example, the following modules might be scheduled when 300 tractors per month are to be produced:

Option:	Four-wheel (25 overplanned)	= 100
	Three-wheel (50 overplanned)	= 275
	Option total	= 375
Three-wheel suboption:	Single front wheel	= 200
	Double front wheel	= 125
	Option total	= 325

The three- and four-wheel options are overplanned by 75 sets of components (375 – 300) and the unique-parts suboptions by 50 more (325 – 275). Except when demand for suboptions is highly variable, overplanning them must also be done very cautiously.

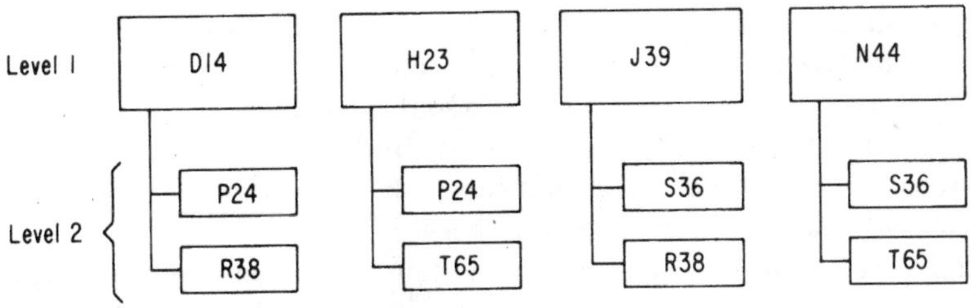

Figure 7-15. A breakdown of unassigned level-1 items.

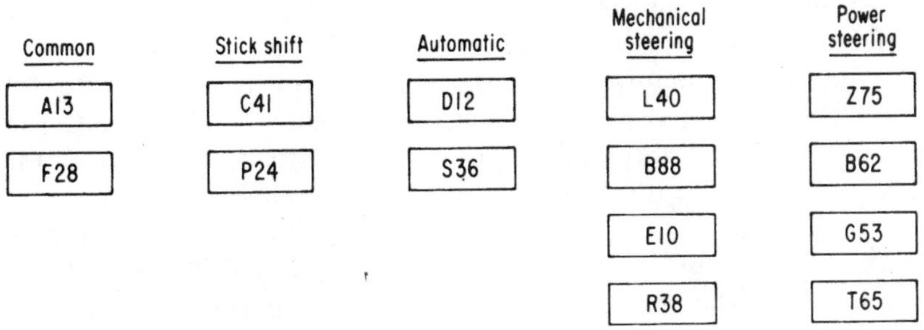

Figure 7-16. Completed modularization.

Setting up a suboption as an option in its own right can pyramid common-parts excess inventory. For example, if the quantities planned for such options were

Four-wheel option = 100

Three-wheel, single-wheel option = 200

Three-wheel, double-wheel option = 125

items common to the three-wheel options would be unnecessarily overplanned by a total of 325 sets compared to only 275 sets in the option/suboption approach. Lack of modularization in product design (called *integrated design*), where option combinations cannot be disentangled and options within options exist, results in higher investment in safety-stock inventory and makes production scheduling poorer.

In this section, two reasons were given for modularization: to disentangle option combinations, and to segregate common from unique components to reduce inventory. In the modularizing example, level-1 items were assigned to different option groups. Many of those items are assemblies and probably contain common components. For example, a subassembly used only with the stick-shift option may have parts common to another subassembly used only with automatic transmissions. Requirements for such common items will be overstated, as they are included in the safety stocks of both options; to avoid this, such parts must be segregated. The option bills have to be further modularized to do this. The additional work and the added files will be worthwhile to reduce inventory *only if there are high-value components.*

BoM modularization is a difficult, tedious task when products are highly complex, are designed as "integrated" assemblies (nonmodular design), and involve many optional features. Judgment is needed in deciding what should be modularized, particularly in attempting to segregate common or semicommon (for instance, an item used with diesel and gasoline fuels but not with LP gas) parts. Excessive modularization should be avoided; if a little is good, a lot is not necessarily better.

Modularization determines the correct level in the original BoM at which to forecast and master-schedule end items. A subassembly promoted to end-item status must appear in the MPS in the proper time-bucket, earlier than the Level-0 products they replace would have appeared, if necessary to allow time for final assembly of the finished product. Each time an item's original bill is modularized and it is transferred to the manufacturing BoM file, such lead time must be recognized. Final assembly and most subassembly lead times are usually short and require only small, if any, time shifts in MPS.

The MPS is part of the planning phase, driving MRP for procurement, fabrication, and subassembly of components. The principal objective of BoM in MRP programs is to define component items required to produce the end items in the MPS. Final assembly schedules are part of the execution process, defining the components needed to build what customers have ordered. When a product BoM is being modularized, each assembly, subassembly, or module is assigned to one or the other of these two schedules:

1. To the master production schedule, by making it part of the planning BoM
2. To the final assembly schedule, by putting it in the manufacturing BoM when needed for execution

Two objectives of modularization brought out earlier were to improve planning by disentangling combinations of options and to reduce inventories by segregating common from unique components. There is a third, broader one: to improve flexibility of production with a minimum investment in component materials inventory, while simultaneously offering a wide choice of products and giving better delivery service to customers. Modular bills of material help to achieve all three of these goals.

Pseudobills of Material

Planning and scheduling components of complex products with a multitude of possible combinations of options requires a very large number of bills of material, 6912 in the tractor example. Modularization reduced this number to 25, but many products can have a much larger number of items to forecast, master-schedule, and maintain BoMs. An accepted principle now is

Forecasts and master schedules improve inversely as the number of end items involved.

A sound planning objective always is to have the smallest possible number of items to forecast and to master-schedule. A technique of creating *pseudobills of material,* also called *superbills* or *S-bills,* helps to meet this objective. Figure 7-16 shows two or more modules associated with each individual option; these could be combined in pseudobills, assigning an artificial parent identity, as illustrated in Fig. 7-17.

Pseudo-, super-, or S-bills are artificial entities created solely to facilitate planning by reducing the number of items to be planned; the items they represent will never be assembled. With S-bills, forecasting and master-scheduling the transmission option in the tractor example in-

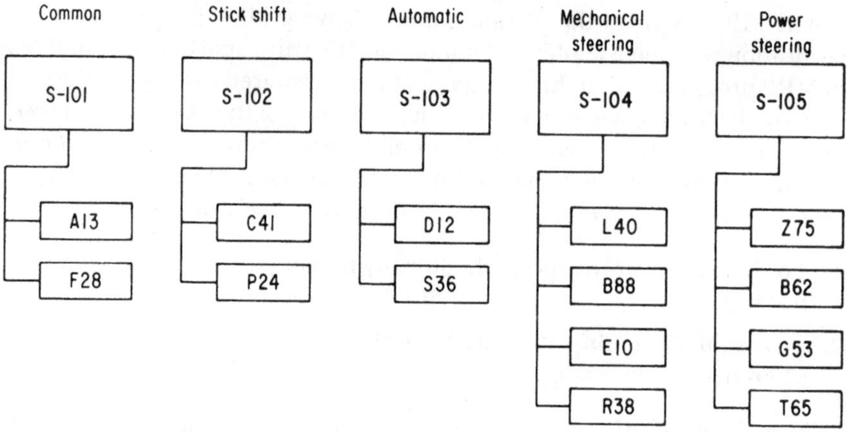

Figure 7-17. Superbills.

volves only S-102 and S-103. These pseudobill numbers then represent this option in the master production schedule, and MRP will explode the MPS requirements utilizing the S-bills in the BoM file.

Another useful pseudobill is the *kit number* or *K-number*, useful where there are many small, loose parts on level 1 in the product structure. These are usually fasteners, other hardware, and packaging materials used in assembling major product units. MRP users find it impractical to deal with such items individually on the MPS level. K-bills put these parts into an imaginary bag, with a part number to identify it; it is then treated as an end item for master scheduling.

The objective with K-bills and S-bills is the same: to assign a single new identity code to sets of items that constitute a logical planning group and, employing the format of a BoM, to relate such items to an MPS requirement. K-numbers may be used to advantage within a modular BoM to streamline material requisitioning, or they may replace the modular bill.

Artificial BoM like option modules, S-bills, and K-bills are intended for use only in the planning phase to improve forecasting, master-scheduling, and MRP. They represent a superstructure in the bill-of-material file that must be maintained along with the rest of this file. This increases the difficulty, labor, and cost of BoM file maintenance, but the extra effort and cost is very worthwhile when they are used properly.

Manufacturing Bills of Material

Level-0 products, those sold to customers, are rarely used in master scheduling. Instead, particularly for complex products with options,

Level-1 and Level-2 modules are established, associated with individual options, and promoted to end-item status in MPS. However, BoM for Level-0 products and other subassemblies excluded from planning BoM are needed in execution; these products have to be ordered, scheduled, and assembled, and requisitions must be issued for components. Final assembly schedules may match customer or warehouse orders in quantities and delivery dates; larger quantities, however, may be assembled and shipped in smaller lots. Marketing, sales, industrial engineering, cost accounting, and others need level-0 BoM also.

Manufacturing bills, called *M-bills,* meet these needs. M-bills are coded to distinguish them from planning bills that MRP uses exclusively. M-bill items can be components of end products or of other M-items. M-bill components may be other M-items or end items in MPS, the top-level items in planning BoM. Purchased and manufactured items procured to support execution of final assembly schedules (rather than to support MPS) belong in M-bills, but they may be included also in kits in planning bills to ensure procurement in time. M-bills are not involved in planning; they are intended for use in execution. *Planning defines the resources needed to support the plans; execution assigns available resources to produce what customers have ordered.*

Ideally, planning should result in available resources being adequate for execution; this never happens. At best, planning is good enough to limit shortages of resources to amounts that can be acquired during execution with minimum cost and little harmful effect on customer deliveries.

The Order Entry Interface

Activities of entering customer orders and managing order backlogs, called *order entry,* are external to MRP except where customer orders or contracts constitute the master production schedule itself (see Chapter 11 for a discussion of potentially serious problems arising from this practice). Order entry is one of the drivers of the execution phase and normally does not interface with MPS but with final assembly scheduling. Allocation tests (see Chapter 9) made of component records to determine shortages, if any, before releasing final assembly orders, utilize planning data maintained by MRP, of course.

Sales literature and other communications with customers employ descriptive English or generic-code model numbers identifying only the principal features of the products. One significant-digit type of generic or product-description code is shown in Table 7-1. Such generic codes are convenient for sales and marketing uses and have the additional advantage of stability, as they are unaffected by model variations or engineering

Table 7-1. Product Description Code

Example: 3 A G 1 1 A P 3

Position	Code	Option
1	1	Model 450 tractor
	2	Model 550 tractor
	3	Model 650 tractor
2	A	4-wheel construction
	B	3-wheel construction, regular
	C	3-wheel construction, special
3	G	Gasoline
	D	Diesel
4	1	56 hp
	2	68 hp
	3	76 hp
5	1	Stick-shift transmission
	2	Automatic transmission
6	A	Regular axles
	B	High-clearance axles
7	M	Mechanical steering
	P	Power steering
8	1	No power takeoff
	2	Power takeoff, type A
	3	Power takeoff, type B

changes; "power steering" remains power steering for many automobile models, but bills of material for this option in specific models are different and will change over time. To make customer orders understandable to planning, production, and other people, and to enable order entry and other computer programs to interact, an order entry editing process translates customer orders into BoM.

Where modular BoM exist, each generic code has one or more planning-BoM-number counterparts, and the generic-to-specific conversion can be made by using decision tables or *inverted bills of material*, another type of pseudobill. These are illustrated in Fig. 7-18, where the diesel option in Table 7-1 (*D* in position 3) coupled with 56 hp (number 1 in position 4) calls out S-201 engine; the diesel/68 hp combination uses the

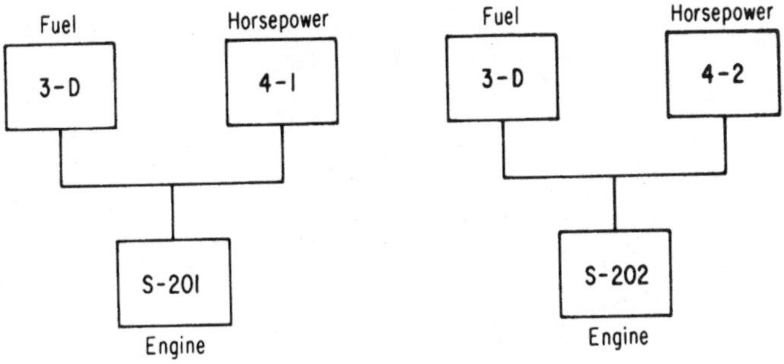

Figure 7-18. Inverted bills of material.

S-202 engine. This links customer order entry, final assembly scheduling, and MRP.

Specifying options during order entry directs bill processor programs to call out appropriate planning BoM and construct M-bills for individual end products. Lower-level subassemblies of these end products will not be called out unless these are part of the M-bill file also. Such linkages are vital to ensuring that plants build, ship, and invoice customers for the specific products they have ordered, and also that the needs of manufacturing planning and control are met. Properly structured bills of material are the key to using MRP effectively in any type of manufacturing. They are of equal importance with valid, realistic master production schedules and accurate data.

PART 3
Applications

You can't solve problems with the same type of thinking that created them.

ALBERT EINSTEIN

8

Developing Valid Inputs

For as you sow, you are like to reap.
SAMUEL BUTLER

Sources of Inputs

All MRP system outputs are produced by processing inputs from five major sources:

1. Master production schedules
2. Orders for components originating from sources external to the plant using the MRP program
3. Forecasts for items subject to independent demand
4. The inventory record (item master) file
5. The bill of material (product-structure) file

All of these inputs are used by MRP to determine the correct inventory and order status of each item under its control. *Master production schedules (MPS)* drive the overall plan of production. They are stated in terms of end items, which may be (shippable) products, highest-level assemblies from which these products are built, or planning modules (see Chapter 7). The span of time MPS cover, termed the *planning horizon,* must be at least as long as the cumulative procurement and manufacturing lead time for all components of the MPS items, if MRP is to provide valid information on them. The planning horizon normally exceeds this cumulative lead time by a significant amount.

MPS are the primary input to MRP, enabling it to translate these schedules into individual component requirements; other inputs merely supply reference data required to do this. If MPS are to define the entire manufacturing program of a plant, they must contain demands for all products the plant will produce, requirements for all components originating from sources external to the plant, and forecasts for all items subject to independent demand. In practice, however, external demands for components and forecasts for independent demands normally are not incorporated in the MPS but are fed directly into MRP as separate inputs.

External demands for components include service-part orders, interplant orders, original equipment manufacturer orders from other manufacturers who use components in their products, and components needed for marketing/sales promotions, research and development, engineering experimentation, destructive testing, and plant and equipment maintenance. MRP treats such needs as additions to gross requirements for components to satisfy MPS.

Forecasts of independent demand for components of MPS items are usually made outside of the MRP program; MRP can be programmed to perform this function, using statistical forecasting techniques, but it is risky to allow such input into MRP without planner review. After reviewing forecasts, planners input, as added gross requirements, quantities they decide are reasonable for such items. Items subject only to independent demand (service parts no longer used in regular production) are better planned using time-phased order points, as described in Chapter 4. Service-part demand must be forecast, unless the service-part organization operates its own MRP or TPOP program and can supply data on the future planned orders it will require.

The *inventory record file,* also called *item master file,* contains inventory status data required by MRP for determining net requirements. This file is kept up-to-date by posting to it transactions reflecting events changing inventory status; such transactions, therefore, constitute indirect input to MRP. In addition to status data, inventory records also contain planning factors used principally for determining the size and timing of planned orders. These factors include item lead times, safety stocks, scrap allowances, and lot-sizing algorithms; they are changed at MRP users' discretion, and such changes normally change inventory status.

The *bill-of-material file,* also called the *product-structure file,* shows the relationships of components and parents that are essential to determination of correct gross and net requirements. It plays a passive role in the requirements computation process, where its function is akin to that of a city directory that the MRP program consults when it needs to call on the inventory records of the components of a parent item. All parents' records carry the storage addresses (called pointers) of the respective bills of material in their inventory records. The inventory file and the bill of material file are thus cross-referenced (chained).

Input Data Integrity

MRP entails processing massive amounts of data; it is a practical impossibility to implement such programs in manufacturing firms without computers, but they perform their programmed functions without regard for the quality of data they process. When these data lack integrity—they are not complete, timely, and accurate—computer-based MRP will fail to provide its users with valid plans. This lack of validity may take some time to become evident and, until it does, people will make poor decisions. When it does, people will ignore the formal MRP output and attempt to develop informal "systems" to replace it. In both cases, the consequences will be far-reaching and severe.

Low-quality input data invariably plague newly-developed MRP programs and continue to hamstring others long after implementation, when disciplines in data handling become slack. Data for use in MRP programs are contributed by people in design and process engineering, planning and control, purchasing, receiving, stockrooms, expediters, dispatchers, truckers, and supervisors, among others; all are able to introduce errors. The quality of input data varies with its source; the incidence of errors is always highest where people view "paperwork" and "record keeping" with disdain, as often occurs in factory and warehouse operations.

Input-data errors cannot be entirely prevented, but means can be installed to detect errors promptly, find the causes, and eliminate them quickly. Detailed discussion of cycle counting and other similar programs is available in the literature.

It is feasible to incorporate into MRP programs a variety of external and internal system checks to minimize the impact of errors. Computer programmers can incorporate many auditing, self-checking, and self-correcting features into MRP. The war against errors should be conducted on three separate fronts:

1. Erection of barriers to keep errors out of the system
2. Detecting internally errors that get through barriers
3. Purging the system of residues of undetected errors

A number of procedures and techniques can serve as *barriers* or *filters*. Computer input audits, testing the correctness of input data (Does such a part number exist? Is this a legitimate transaction code? Are any of the data missing?) are typical, and commonly used before processing begins.

Diagnostic routines can be programmed that will conduct other tests prior to the actual processing of input data. For example, before processing receipts of an item, a diagnostic test against open-order records might indicate that no order has been issued. Diagnostic tests run on files other than those to be updated, or on special tables set up for this purpose, will cost extra processing time; this may be far less than the costs of input-data errors. Modern computers make possible a great variety of this type of check; it is the swiftest and most efficient way of catching errors before bad data are processed.

During actual processing, *internal detection* of errors that got past the barriers can also be programmed on computers. These are distinguished from diagnostic tests by being made as files are updated. An example is checking a stock-disbursement transaction that has already passed both input audit and diagnostic tests, and detecting an error in a withdrawal quantity that exceeds the quantity previously on hand.

A test for "reasonableness" can also be employed. For example, if usage of an item has averaged 100 per period, a gross requirement of 1000 in a period is highly suspect. Unit-of-measure errors are common; liquids ordered in gallons are reported as received in (42-gallon) barrels. A person will spot and question such absurdities immediately; a computer can be programmed to do the same. Computer programs applying reasonableness tests usually reject and "flag" suspect results, telling output recipients not to enter the information before verifying it.

Purging residues of errors that have escaped detection is mandatory if MRP data are to be kept from gradually deteriorating. A small number of errors in input data will reach MRP files despite all barriers and checks. These errors may be almost impossible to stop, but procedures can be devised to detect and eliminate their harmful effects on MRP outputs.

This is accomplished by various reconciliation, purging, and close-out procedures, which are analogous to writing off periodically miscellaneous small unpaid balances in an accounts-receivable file. An example of this type of procedure is closing out old shop or purchase orders still showing small quantities due.

Most MRP programs running in industry contain some percentage of error at all times. Small percentages can be tolerated as long as error detection and purging prevent system-resident errors from accumulating continually. Accumulation at even a very low rate will eventually hamstring MRP. When clearing out error residues, it is not so important how soon after the error occurred it is removed, but it is all-important that it be removed at some scheduled interval. Some level of residual error is inherent in any working MRP data, but it must be kept to a minimum.

Even more important than the technical aspects of preserving data integrity are the training, discipline, and attitude of the people who handle and use the data. Those who contribute input data and those who use MRP outputs in the performance of their jobs must understand that a computer's outputs cannot be better than its inputs. If MRP is to be used successfully and constantly rising levels of data accuracy are to be achieved, management must accept responsibility for the training, discipline, and motivation of everyone who handles data. Total Quality Management applies also to the quality of all data used in the business.

Master Production Schedules

Master production schedules make up one of the three principal inputs to MRP. The other two, inventory status and product structure (see Chapter 7 for detailed discussions of these), simply supply reference data; MPS are the inputs that "drive" MRP. It is the first step in the implementation of the overall manufacturing program of a plant, represented by data in MPS. These data are the wellhead of the flow of manufacturing logistics planning information. MRP depends on the validity and realism of the MPS for its effectiveness.

What is commonly called "the master production schedule" is really a set of individual schedules for each item planned at this level. Each individual MPS has a supporting bill of material defining the components needed to produce the planned quantities of the end item; the aggregate of all master production schedules determines total future loads and inventory investment.

The objectives of master scheduling are to

1. Develop data to drive detailed planning
2. Provide devices to reconcile customers' desires with plant capabilities
3. Furnish means to make reliable delivery promises and to evaluate the effects of schedule changes
4. Coordinate plans and actions of all organization functions and to measure their performance
5. Provide management with means to authorize and control all resources needed to support integrated plans

Considerations in developing and updating MPS, although not actually part of MPS data, include external (customers) and internal (service parts, warehouses) demand forecasts, customers' orders, new product and market opportunities, major inventory changes, details of processing and plant facilities, and suppliers' capabilities. Other factors such as capital availability may become serious considerations at different times.

The validity of these MPS establishes how well managements' policies and goals will be met; their realism determines how well the plans will be executed, and what level of performance will be reached in customer service and profitability. Downstream systems are unable to compensate for deficiencies of their input. MRP will carry out its functions of priority planning and provide data for capacity requirements planning with great efficacy; it will provide useful data only if it is presented with realistic, valid master production schedules to be processed.

Functions of Master Production Schedules

MPS have two principal functions:

1. In the short horizon, to serve as bases for planning material requirements, production of components, order priorities, and short-term capacity requirements
2. Over the long horizon, to serve as bases for estimating long-term demands on company resources such as people, plant, and equipment, warehousing, and capital.

The first function relates to the "firm" (front end) and the second to the full horizon of MPS. The firm portion, usually equal to the sum of production lead times from release of purchase requisitions to completion of orders for MPS-level items, can affect many open purchase and work orders if MPS data are changed. The portion beyond involves only planning data and computer files (plus an occasional very-long-lead-time item).

Designations of "firm," "flexible," and "unrestricted" are applied to portions of MPS. *Firm* recognizes the desire to avoid MPS changes affecting open orders; in this portion, a single MPS change can alter requirements for dozens of next-lower-level components, hundreds more one level below that, and thousands more at lower levels. All such items may not have open orders, but many will. *Flexible* recognizes that some changes can be accepted in the mid-portion of MPS, but that resources may not be available for extreme differences. The *unrestricted* portion has no constraints.

"Nervousness" resulting from changes in the firm portion is caused most often by users' attempts to substitute customers' orders for MPS items, thus failing to distinguish clearly between planning and execution, discussed in Chapter 10. This is self-defeating; MRP cannot provide the same precision of data on actions needed to overcome shortages and ship products that had not been included in MPS. How allocations are used to do this is covered in Chapter 9.

A well-implemented MRP program utilizes the entire MPS planning horizon in its time-phased inventory records. MRP thus maintains data on all planned requirements and planned and released orders, to provide visibility into the future item-by-item. These data are used in a variety of ways; among them are setting lot sizes; determining capacity requirements, inventory investment, and cashflow; showing requirements for negotiation of blanket-order contracts with suppliers; and determining inventory obsolescence and the resulting write-offs.

Preparing Master Production Schedules

MPS represent future demands on production resources resulting from planned output. The method of establishing these demands varies with the type of business. In the manufacture of products to stock, future requirements generally are derived from forecasts based often on past demand. In manufacturing to order, customer orders on hand may represent the bulk of total production requirements. In custom assembly of standard components, known as *assemble-to-order,* a mixture of forecasts and customer orders generates requirements. The organization of the distribution network and field warehouse policies also directly affect production requirements.

Requirements in most manufacturing companies derive from several sources. Identification of these sources and of the demand they generate constitutes the first step in developing MPS. These sources include

1. Customer orders
2. Dealer orders
3. Finished goods warehouse requirements
4. Service-part requirements
5. Forecasts
6. Safety stock
7. Orders for stock (stabilization inventory)
8. Interplant orders

Customer orders may constitute MPS with simple products, custom-engineered products, contract manufacturing for the military, true "job shop" (make-to-order tools, dies, and test equipment), and in rare situations where order backlogs extend beyond the cumulative production cycle time. As discussed in more detail in Chapter 11, the distinction between customer orders representing demand on product inventories and MPS representing supply of needed items must be kept clearly in mind, particularly if plants are operating behind schedule on deliveries.

Dealer and warehouse requirements for products may be treated in MPS preparation the same way as customer orders, if data beyond the next order are provided. Time-phased order point and MRP programs in dealer and warehouse planning systems are often linked directly to supplier MPS to provide such advance data for specific products (see Chapter 11 for further comment on this). If advance commitments are stated in general terms of product models without specific choices of optional features, the supplier plant must forecast the options when developing its MPS.

Service-part requirements of customers or service warehouses normally bypass MPS and are entered as forecasts or orders directly into the items' gross requirements data. Large, expensive service-part assemblies should be entered in MPS along with regular products.

Forecasts may be the sole source of requirements for plants that ship directly to customers from a factory warehouse or that assemble products to order. In other businesses, forecasts are large portions of total demands to be input into MPS. Specific product variations or optional features usually are forecast by plant people, even though MPS and component production are based on product units ordered by dealers or field warehouses. For final assembly scheduling, exact configurations of optional features are supplied closer to shipment dates.

If deemed desirable, *safety stock* in MRP programs, as mentioned in Chapter 6, are better planned on the MPS level rather than on component levels, and therefore must be viewed as a separate source of demand on the plant.

Orders for stabilization stock may be the principal source of production requirements where products are being stockpiled in anticipation of future needs. In businesses subject to seasonal demand, products and/or components normally are produced to stock during the off-season to be able to meet peak demands while maintaining level loads on productive facilities.

Interplant orders normally cover component items rather than products, and may include anything from single-component parts to (assembled) end items appearing in MPS and are handled like service parts. Where the "customer" plant uses MRP, such demand is more effectively conveyed to supplier plants using planned-order schedules from the customers' MRP program.

Demands from all these sources, when consolidated, represent total factory requirements. The creation of these schedules constitutes the second step in the development of MPS. This is derived from the first but is not necessarily identical to it, for the following reasons:

1. A part of demand on the factory may be met from plant inventories.
2. Product lot sizes, important to manufacturing, are not reflected in schedules of factory requirements. In developing MPS, product lot sizes are established that may deviate, in both quantity and timing, from source requirements. Additional lot sizing may subsequently take place at component-item levels.
3. The load represented by the schedule of factory requirements may exceed or be less than the productive capacity to which the plant is committed.
4. This load may fluctuate excessively.
5. The schedule of factory requirements may show product models, and will have to be translated into end-item BoM numbers. It may not specify optional product features, the demands for which must be forecast before being incorporated into MPS.

Final preparation of MPS, based on the schedule of factory requirements, is the third and final step, creating a specific manufacturing program to be processed by MRP to plan component procurement, fabrication, and subassembly activities. In transforming the schedule of factory requirements into MPS, the predominant considerations are factory data needs and capacity availability in the present and future.

The logic of MPS development and implementation activities is illustrated in Fig. 8-1. Both phases have closed-loop feedback to revise MPS data if resources, including time, are not available and, therefore, plans would be unrealistic.

Making Master Schedules Realistic

Managing inventories and production properly requires valid and realistic MPS. *Valid* means that they support management's strategies and policies; *realistic* means that resources needed for execution will be available. MPS must be constrained by limits to available productive capacity of suppliers and company facilities over the short horizon. When

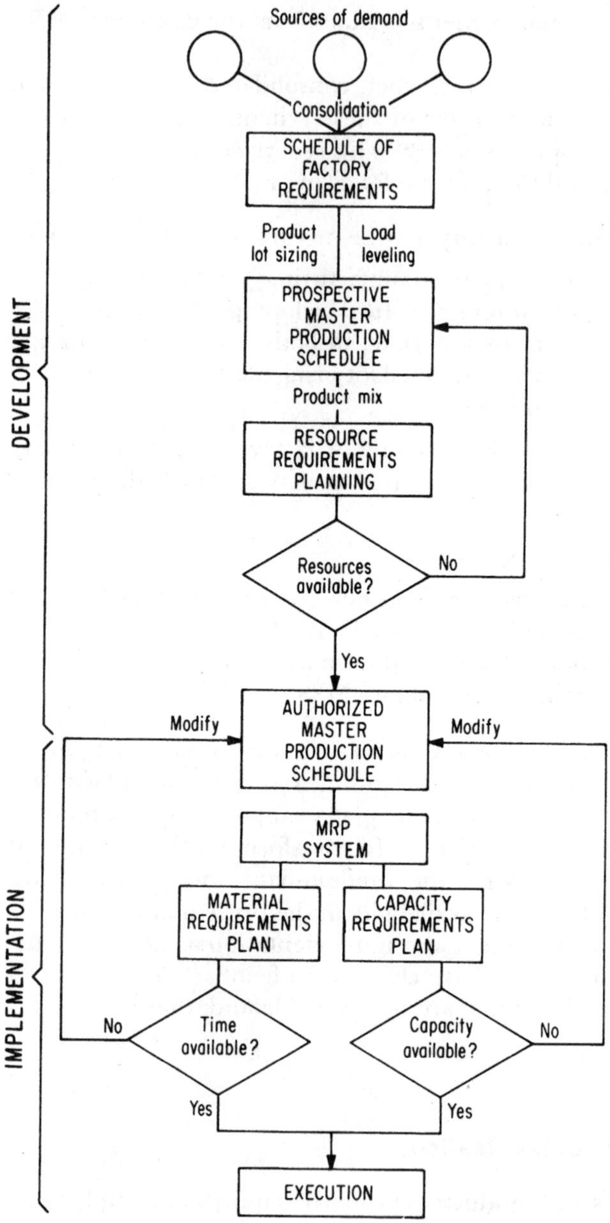

Figure 8-1. Master production schedule development and implementation.

disparities develop between MPS requirements and the capacities of plants and their suppliers, every effort must be made to reconcile the two. The usual first step is to determine if some extraordinary actions will provide resources adequate to execute MPS; overtime, subcontracting, and alternate operations can often overcome significant capacity shortages.

More serious and difficult problems arise if such actions are inadequate and MPS cannot be produced; there is then no real alternative to revising MPS promptly. The tough question is, "Exactly what in which MPS will be changed, and how?" MRP can help with the answer. Pegged requirements (discussed in Chapter 3) permit tracing upward through BOM from specific components affected by a production problem to end-item MPS.

Some problems may be solved below the MPS level by revising planned-order data in parent-item inventory records, as shown in Chapter 4. Pegging permits tracing to any higher level as necessary, to pinpoint specific lots needing to be changed to restore harmony between MPS and reality of resources.

MPS provide data for determining, over the long horizon, plant workers and machinery capacity requirements as well as estimates of other resources (tooling, energy, utilities) required to execute them. Some resources, such as plant and new machinery, may take a year or more to acquire; while MPS data may look more precise than rough-cut means for estimating such requirements, they may not be as accurate. This is discussed later in this chapter.

Management and the Master Production Schedule

It is sometimes suggested by computer experts working on networked manufacturing logistics systems that MPS should be automated under complete computer control. As mentioned earlier, use of statistical forecasting of demand tempts underqualified people lacking sound understanding of how manufacturing can and should work to integrate the MPS and MRP programs. They even suggest that planner intervention in order release should be eliminated and all MRP lot sizing, explosions, netting, and rescheduling orders be computer-controlled. Their reasoning sounds convincing: the logic of the procedures can be clearly defined, and all the required data are available.

Unfortunately, all the required data are *not* available. Who can predict what upsets will occur and what changes to plans will be desirable or mandatory? Planning will never be perfect, and replanning cannot substitute for knowledgeable execution reacting to unexpected changes. Information on a multitude of extraneous factors—company policy, markets, competitors' actions, government regulations, new technologies—cannot be programmed into computer systems; reacting properly requires seasoned managerial judgment. Computers can remove the "coolie labor" of clerical data-handling very cost-effectively; what they cannot do is replace human powers of analysis and decision making.

Master production schedules properly can be called "managements' handles on the business." They are planning devices, not execution tools. They are statements of what should and can be made, balancing customers' needs against manufacturing capabilities. They provide means for coordinating the activities of all functions and measuring their performance, as well as giving management control over manufacturing resources. MPS are not sales forecasts, customer orders, or assembly, packaging, and shipping schedules.

Coupled with modern MRP programs, MPS constitute means for the solution of many problems often neglected in manufacturing. To realize their full potential benefits, users must understand fully the necessity of maintaining close correspondence between MPS and realities on the manufacturing floor. The key to this is managements' willingness to make MPS realistic.

This requires a departure from the traditional views of top-level manufacturing plans as "challenging goals driving plants to greater efforts" and "contracts with customers that cannot be changed." This chapter stresses the need for realistic MPS, subject to continuous review and adjustment—a living plan.

Managements' role with MPS driving modern MRP should be to

1. Understand their function of linking company policies and goals with detailed plans
2. Set policies and guidelines for all functional groups involved with MPS, and measure their performance
3. Insist that MPS be realistic
4. Resolve conflicts among user groups
5. Review MPS regularly and rigorously

The alternative to sound management understanding and close involvement in master scheduling is resource planning that is driven by invalid data. This is the single most common reason expensive MRP programs have failed to help manufacturing companies achieve the enormous potential benefits.

Lead Time

One of the most important inputs to MRP programs, affecting much more than order start and due dates, is lead time. The lead time of a manufactured item is made up of five elements, listed here in descending order of significance:

1. Queue time (waiting to be worked on)
2. Running time (processing the batch)
3. Setup time (changing equipment from item to item)
4. Move time (including waiting to move to next operation)
5. Preproduction time (preparation to release orders)

Preproduction time includes time for all paperwork and related activities needed prior to releasing orders; examples include preparing requisitions and placing purchase orders with suppliers, and making up the shop packet authorizing work. It is *unnecessary to include preproduction time in lead times in MRP*; the forward visibility of planned order release dates allows these activities to be done before these dates become current.

Move time is a function of plant layout and material handling activities. Modern plants utilize machine cells or work teams making families of similar items and passing them, often by hand, to successive operations, making move times negligible. Even where heavy or bulky materials are handled, requiring fork trucks or mechanical transfer, move times can be kept short by using due-date priorities to move urgent orders promptly.

Setup time is a critical element of lead time, although usually it is not a large factor. It determines order quantities (through calculations discussed in Chapter 6) and causes equipment capacity losses during changeovers. Order quantities establish the work load on processing facilities and influence the flow of work being processed. Large orders, lumps of work, cause large, erratic queues of work-in-process; small orders flow more smoothly and require only small queues to avoid running out of work. Short setup times make short queues possible.

Running time is rarely a major element of lead time. It is determined by the processing methods and the size of orders. The latter is related to setup time, as just explained.

In poorly planned and controlled plant environments, *queue time* accounts for up to 95 percent of average total lead time; well-run plants have 20 percent and less. Queue time is the result of an amount of work at a work center awaiting processing; each order in the queue will experience a delay, depending on its priority relative to that of other jobs. The queue time of any order, and consequently its actual lead time, will increase or decrease as its priority is changed; the "hottest" orders will spend little time in queue. Actual lead times in emergency situations can be compressed to a small fraction of the planned lead time.

Due dates are the primary factor (the amount of processing work is sometimes considered) used in setting work priorities; the nearer the due date, or the longer past due, the higher an order's priority. MRP has the ability to reevaluate open-order due dates based on the latest data fed to it for replanning.

Planned lead times can be computed by detailed analyses of setup and processing work standards, in-plant travel distances, and actual queue times, but the precision thus achieved is not worth the effort. Planned lead times are averages needed by MRP to relate order release times to completion times; actual order lead times are manageable through priority control.

Planned lead time is sometimes artificially inflated by the inclusion of an element called *safety lead time* or safety time. This element is inserted at the end of the normal lead time; order due dates are thus advanced from planned due dates by the safety lead time. MRP offsetting for lead time will then plan order release earlier, also by the amount of safety time.

Safety lead time acts similarly to safety stock; both attempt to compensate for the vagaries of item demand. If the plan is executed well, safety lead time creates an inventory excess that acts as a cushion against unanticipated demand. MRP logic attempts to protect safety stock and time from being used, as discussed in Chapter 6, adding to cushions. In practice, however, knowledge of the planned cushions causes people to pay less attention to MRP planned due dates, and this extra inventory remains in work-in-process. This inflates queues, lengthens lead times, and results in more late orders.

Resource Requirements Planning

A very common and popular acronym is MRPII, standing for Manufacturing Resource Planning; the II indicates a broader meaning than MRP (material requirements planning). It was applied originally to the core system described in Chapter 1 (which includes both capacity requirements planning and control and priority requirements planning and control), together with engineering, marketing, and cost data-handling programs. The acronym was an unfortunate choice, causing confusion and generating more heat than light on the subject of improving operations. In this book, the term *resource requirements planning* refers strictly to capacity planning for resources used in manufacturing.

Master production schedules must be realistic—capable of being produced. Current available and future planned resources including capacity, space, and working capital, must be adequate. The problems arising when available resources are not adequate to support MPS are extremely serious; this topic is covered in Chapters 8 and 11. Overstated MPS cause poor customer service, excessive and unbalanced inventories, and high costs—the worst of all manufacturing worlds.

Resource requirements planning over the long horizon provides data to test the validity of Production Plans (see Chapter 1) and MPS. The objective is to balance customers' desires against plants' ability to meet them, with reasonably level loads on resources. Five steps are involved:

1. Defining the resources to be considered
2. Developing load profiles for products or families of products showing the loads imposed on resources
3. Extending these profiles by the quantities called for by proposed Production Plans and MPS to determine the total load on each resource in question
4. Simulating the effect of alternative Production Plans and MPS for improving performance and/or contingency planning
5. Selecting realistic MPS that make the best use of existing or planned resources

Defining the resources to be considered in MPS development is a management function. Typical resources included are:

1. People—direct and indirect labor and support people with special skills (such as design and industrial engineers)
2. Materials—purchased and manufactured
3. Tooling and test equipment
4. Plant and equipment
5. Capital
6. Energy
7. Utilities—power, water, sewers, and environmental pollution control equipment

Resource planning may cover large groups such as an entire machining department, subgroups of similar machines (lathes, presses, automatic insertion, etc.), or, rarely, individual critical machines. Its purpose is not to determine exact loads on specific resources but rather to evaluate the overall impact of Production Plans and MPS. It is conducted on a "macro" level using rough approximations of load; a precise fit is not sought. The objective is to develop load data quickly and with reasonable accuracy, so that alternatives may be tested easily. Future uncertainty in planning makes precise data of questionable value.

Developing load profiles is rarely done for individual products unless these generate a very significant amount of measurable load. In manufacturing large generators or turbines where this is true, the same procedures used to determine machine loads are applied to compute product load profiles. These show standard hours required by period to produce one unit of product in whatever fabrication resource is selected. Fig. 8-2 is a load profile for total fabrication of a product appearing in a master production schedule in Period 10; the overall fabrication load it would generate is distributed over eight preceding periods (the fabrication lead time).

Extending load profiles for individual products by the quantities in its MPS, and summarizing them by period, is simple and quick in a computer able to access load profiles. The result is a printed report or visual display of loads, called *resource requirement profiles,* on the various resources studied over the planning horizon. Loads may be segregated by individual product lots, as illustrated in Fig. 8-3, to show which of these may cause capacity problems.

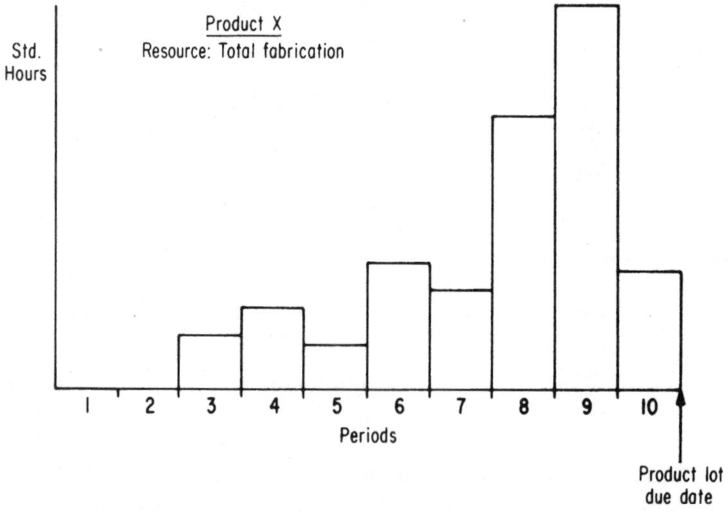

Figure 8-2. A product load profile for total fabrication.

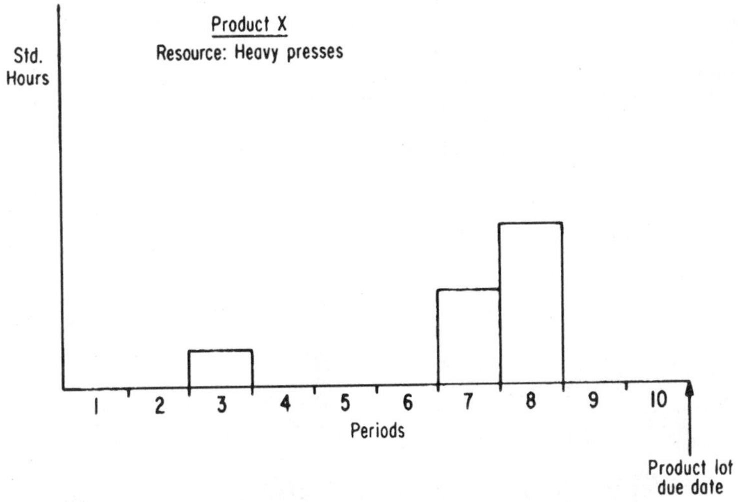

Figure 8-3. A resource requirement profile.

In the figure, loads generated by service-part and interplant requirements have been added to those from product lots. These could be a percentage of the product load based on past experience or forecasts, or computed through separate load profiles if service-part and/or interplant-item demands are large.

Load and *capacity*, terms often used interchangeably, have very different meanings. Load is *an amount of work at a work center at any time*; capacity is *the ability to process an amount of work during a time period*. Load is the depth of water in a tank at some point in time; capacity is the rate it is running in (input) and out (output) of the tank during the time period. To translate load into capacity requirements, corrections have to be made for loads planned as queues of work-in-process needed to prevent starvation and consequent loss of output because of erratic work flow. How to do this is described in Chapter 9.

Most manufacturing operations involve many products and work centers. Detailed capacity requirements planning using specific MPS data, BoM, and operation standard hours over long planning periods look highly precise but are not accurate enough to justify the massive data handling required. Instead, the family groupings and data in the Production Plan are used to produce a *rough-cut capacity requirements plan*.

The technique begins with bills of labor (BoL) (see Fig. 8-4) for *families of similar products*; these are grouped by processing operations in work centers. Hour estimates are based on standards, if available, or previous experience, and cover all work on family and components. Estimates are made of total standard hours required to produce one unit of each family

Work center	Product family #1, estim. hours	Product family #2, estim. hours
101	—	—
102	10	9
103	—	—
104	8	12
105	—	—
106	14	18
107	—	—
108	22	20
109	—	—
110	—	—

Figure 8-4. Bills of labor.

in each work center involved. The name *bills of labor* (*BoL*) comes from the similarity to BoM; making one family (parent) takes *x* hours (pieces) in one work center (component). Common bill-of-material processor programs will load bills of labor into computer files, exactly like BoM. Supplier capacity can be included also by adding to BoL the proper units (pieces for fasteners, molds for castings, gallons for liquids) that they use in measuring capacity.

The following example illustrates rough-cut capacity planning. Production Plans for two families, units per month, are

	Jan	Feb	Mar	Apr	May	Jun
Family #1:	120	130	140	140	150	150
Family #2:	55	55	55	55	55	55

Extending these quantities by the hours in BoL in Fig. 8-4 gives one month's work-center capacity requirements, as shown in Fig. 8-5, for the two families. This is true capacity—throughput—not load data. Capacity requirements for all families over additional months are shown in Fig. 8-6, together with current demonstrated capacity based on recent throughput. Underloaded (W.C. 105) and overloaded (W.C. 106) centers are immediately evident, and corrective actions can be initiated.

Tracking actual queues of work at work centers will monitor BoL hour estimates for errors. Understated estimates will cause queues to shrink, and vice versa, so errors are soon evident.

The technique tests both MPS and Production Plan validity, if MPS quantities for families of products equal Production Plan family totals as

Work center	Product families #1	#2	—	—	—	Jan. total
101	—	—				3,210
102	1,200	495				7,495
—	—	—				—
106	1,680	990				9,220
107	—	—				5,425
108	2,640	1,100				13,455

Figure 8-5. One month's capacity requirements.

they should. Families having widely different models are subject to serious errors in BoL hours required if the mix of models within a family changes significantly. Care should be taken to avoid this or, at the least, to monitor the mix frequently.

Rough-cut capacity requirements planning both simplifies and improves capacity planning. It is the only way to keep planning up-to-date with design and processing methods changes, increasing in frequency in global competition. It is often criticized for ignoring lead time offsets, lot sizing, and components in stock; it assumes that needed quantities of components to make the Production Plan totals will be produced each month. The effects of lead time offsets and lot sizes vanish if several month's data are averaged. Unless component stocks are to be reduced significantly, those used will be replaced and no error will result.

Bills of labor can be expanded into *bills of resources* (*BoR*) by adding estimates of requirements of other resources as shown in Fig. 8-7. Data may be criticized as approximate, but can be made *accurate enough for all practical purposes.*

Load profiles, bills of labor, and bills of resources may be stored to be used in resource requirements planning without repeated detailed com-

Work center	Demon. capac.	Capacity requirements Jan.	Feb.	Mar.	Apr.
101	3,300	3,210	3,290	3,310	3,310
102	7,540	7,495	7,650	7,725	7,860
105	7,750	6,630	6,770	6,900	7,070
106	8,300	9,220	9,220	9,220	9,220
107	5,650	5,425	5,425	5,425	5,425
108	13,535	13,455	13,620	13,790	13,945

Figure 8-6. Capacity requirements plan.

Product family 6A31	
Work center 102:	10 hrs
Work center 105:	8 hrs
Work center 106:	14 hrs
Work center 108:	22 hrs
Work center 108:	22 hrs
Work center 108:	22 hrs
Tool maintenance:	21 hr/100
Testing:	3 hr/100
Energy—Electricity:	12 kwh/100
Water—Process:	35 gal/100

Figure 8-7. Bill of resources.

putations, unless products are redesigned significantly; most engineering changes have relatively small effects on loads. The extra work of loading and maintaining BoL is well repaid by better capacity requirements data.

Simulating the effect of MPS alternatives starts the selection-decision process. If proposed plans generate significant over- or underloads in one or more periods, alternative schedules are studied until a satisfactory answer has been obtained. This requires knowledge of plant and supplier facilities and judgment. In the absence of such a procedure, MRP will process all MPS, and loads calculated subsequently may be undesirable. MRP will have to be rerun with MPS changes, possibly several times, until a satisfactory load results. The use of load profiles avoids such problems.

Selecting feasible MPS is the final step in this process of ensuring that MPS driving the materials plan are realistic. Adjustments will be made subsequently in short-range capacity requirements planning to determine need for overtime, work transfer, or subcontracting to handle small load fluctuations during execution of the plan. In the typical manufacturing firm, MPS agreed to by management correspond to some specific rate, or level, of production (60 machines a month, 80 vehicles a day) to which all activities are then geared. Just-in-Time advanced manufacturing practices use such rates as MPS to drive MRP programs and as a base for resource requirements planning.

Organizing for MRP

Master production schedules document a company's overall manufacturing program. The development and administration of these vital data should be the joint responsibility of the four basic functional divisions: marketing, finance, manufacturing, and engineering. The first three are involved continually and engineering intermittently, when major redesigns or the introduction of new products affects the manufacturing program.

The responsibilities of the four divisions relative to the manufacturing program should be as follows.

Marketing

1. Forecasting customer demand, answering questions about what products can be sold, how many, and when
2. Setting finished goods inventory levels, in units, model mix and storage locations, unless this responsibility is given to a distribution organization

Finance

1. Financing and budgeting raw material, work-in-process, and finished product inventories
2. Providing cost data for decision making

Manufacturing

1. Development of master production schedules within the constraints established by management
2. Performance to realistic master production schedules

Engineering

1. Designing not only for product function but also for manufacturability
2. Working cooperatively with users of BoM to meet their needs for information and to reduce engineering changes

Finance usually is not concerned with production details, but works with the dollar equivalents in fiscal periods. It must become less occupied with keeping better scores and more with helping the players to make better operating decisions. This requires substantial revisions in standard charts of accounts.

Marketing, like manufacturing, deals in specific units of product, although commonly using larger time periods in forecasts and sales statistics and more precise data in actual deliveries. Preoccupied with the day-to-day problems of selling more, they need to become more aware of the effects on manufacturing of customer changes and new product introductions. They must take responsibility for executing agreed-upon plans for both product mix and sales volume.

Manufacturing must become less obsessed with supposedly rigid production needs and apparently insoluble problems and develop flexible operations to respond quickly to market changes and customer desires.

Engineering, in collaboration with processing, planning, purchasing, quality management, and production, called "concurrent engineering," can develop designs with fewer components (hence lower material and assembly costs), more modularity and common components (for better planning), and products becoming specific models at final assembly (giving greater flexibility).

Successful operations in highly competitive global markets demand that functional divisions change from all-stars to tightly knit teams of fully qualified people who know how manufacturing should work and who pull together to make it work that way.

Specific responsibilities in three areas need clarifying:

1. Forecasting vs. master scheduling
2. Holding inventories of components vs. finished products
3. Providing optional product features

Forecasting demand is clearly a marketing/sales responsibility, whereas scheduling production is (or should be) manufacturing's. Forecasts and MPS are sets of numbers for two different purposes; the former estimates external demands and the latter specifies internal production to meet them. In practice they often are confused, and raw forecasts are allowed to be MPS. When statistical forecasting on computers was introduced, well-meaning but misguided MRP software designers linked forecasts to MPS directly, usually with serious problems resulting. The best forecasts are made by teams of marketing/sales, planning, and production people, reviewed in aggregate by top management.

The authority for specifying and changing MPS is often improperly assigned or not clearly assigned. Marketing and sales like to exercise influence over master schedulers to get desirable changes made in MPS, without understanding or caring about the consequences to manufacturing. Teamwork avoids this oversight.

The policies and actions of all four functional divisions have strong effects on inventories. Some firms, especially those making consumer products, split responsibility for inventories. Manufacturing is accountable for plant inventories of raw materials, work-in-process, and finished components made to support MPS; marketing/sales controls and accounts for finished products in plant and field warehouses and dealers stocks.

This can have dire consequences if marketing/sales are allowed to order products unilaterally without considering the effects on production.

Responsibility for component materials for options is sometimes divided between marketing and manufacturing, based on which is better able to determine future demands. This is possible by ranking options by percentages of the total cost of the product. Above some arbitrary percentage, marketing can concentrate on and forecast these important options; manufacturing handles the balance, using statistical analyses of past demand as forecasts.

MRP is part of the total logistics system that manufacturing companies need to regulate the flow of materials through the entire cycle from suppliers through plants to customers. As mentioned in Chapter 1, it is a principal element in the core program in such systems, improving coordination of data in an integrated plan. MPS drive MRP and provide the mechanism to coordinate the activities of functional divisions and to resolve the inevitable conflicts between them. MPS are often considered "contracts between them," but this leads to rigidity instead of mutual efforts to meet customer demands and company goals simultaneously.

Master scheduling is best when done by teams of people representing marketing, engineering, manufacturing, and finance, and when reviewed by top-level management. The creation and maintenance of MPS are too important to be entrusted to one (probably biased) functional division of any company.

The Master Scheduler

Master production schedules are a major input to MRP; in fact, they drive this important technique. The role of master schedulers, individuals who oversee the development and use of MPS in manufacturing planning and control, is of special importance. Master schedulers have the following duties; they

1. Compare actual and forecasted demand, and suggest revisions to forecasts and MPS
2. Convert forecasts and order-entry data into MPS
3. Correlate MPS with shipping and inventory budgets, marketing programs, and management policies
4. Maintain MPS data files
5. Participate in MPS meetings, preparing agendas, anticipating problems, providing data for possible solutions, and bringing conflicts to the surface

6. Evaluate suggested MPS revisions
7. Develop and monitor customer delivery promises

Tracking actual demands and comparing them to forecasts will provide early warning signals that MPS revisions are necessary. Master production schedulers *convert forecasts and order-entry data into MPS,* changing general descriptions and generic model numbers on customers orders into specific end-item bill-of-material numbers, dividing monthly into weekly quantities and applying forecasts of product options.

They *check MPS data with budgets* for shipments and inventory to detect significant differences for management review. They track actual demand against marketing programs and compare actual performance with management policies of customer service, inventory investment, and cost control.

They *maintain MPS data* by tracking the use of safety stock provided at the master production schedule level, accounting for differences between quantities of end items produced and those consumed by the final assembly schedule, and enter and edit all changes to MPS files.

Master schedulers *play important roles in production meetings,* preparing the agendas, evaluating problems brought to their attention by factory supervisors, inventory planners, and others detecting them, anticipating suggested solutions and having relevant data available to avoid wasted decision-making time, and bringing out conflicts among people to avoid their being ignored, concealed, and neglected.

They *review suggested MPS changes,* determine which end-item lot should be changed and what the effects will be, and initiate recommendations to management on acceptance or rejection. They evaluate the impact of new or changed orders on *customer deliveries* and recommend and monitor promise dates.

The position of master scheduler is broad and diverse. They are the links between many nontechnical and nonrigorous activities and very rigorous, sophisticated, and powerful MRP. Their competence, together with that of planners, determines the usefulness of this vital tool.

Final Assembly Schedules

The distinction between the master production schedule and the final assembly schedule is a source of frequent confusion. The two may be identical numbers, although always different in concept, in some companies where the product line is limited or where the product itself is simple; lawn mowers, hand tools, bicycles, vacuum cleaners, and clocks are examples of products where the shippable product may be the MPS end item. Manufacturers of complex weapons and heavy machinery attempt to make MPS and final assembly schedules identical, and they experience very serious problems. This topic is addressed in Chapter 11.

Between these extremes of very simple and very complex products lie the many products assembled from a few standard components, often to specific customer orders. This category includes automobiles, machine tools, home appliances, electrical equipment, and many others. For these, two distinct schedules are required: MPS for end-item components, and final assembly schedules for shippable products. Because of the disparity between required manufacturing lead times (long) and customer-allowed delivery times (short), MPS must be formulated and MRP employed long before final assembly schedules are prepared.

MPS usually extend several months into the future; the final assembly schedule usually covers only a few days or weeks. MPS often are based on forecast customer demand; final assembly schedules usually contain actual customer orders and may be constrained by shortages of components.

Most components are ordered in support of the MPS in the planning phase; since plans are never perfect and are never executed exactly, shortages can be expected. There are many possible causes of such shortages; a few are forecast errors, BoM and other record errors, late supplier deliveries, machine breakdown, and scrap. Correctly identified prior to final assembly (see "Uses of Allocations" on p. 223 in Chapter 9) many shortages can be overcome quickly and delivery delays avoided at low cost.

As mentioned in Chapter 7, a few selected manufactured and purchased items may be put into M-bills used in the execution phase. They will be manufactured or procured during execution of the final assembly schedule. Such items are characterized by

1. High unit cost
2. Short procurement or manufacturing lead time
3. Short assembly lead time of their parent, if any
4. Absence of long setups or quantity discounts

Here is a make-part example. A horizontal milling machine has a component called an *overarm* that is required in the fourth week of final assembly. The overarm is essentially a steel cylinder requiring little machining and a minor setup, but it is a massive and relatively expensive part. It is assembled by inserting it into a hole in the column and fastening it inside. Different overarms are used in different models.

Assigning this to the final assembly schedules permits fast machining of small quantities during final assembly of specific machines. This eliminates the need to forecast the item, carry it in planning and execution BoM, and commit expensive materials to specific components long before true requirements are known.

Purchased parts often are handled better in final assembly schedules. For example, tractor rear tires are very expensive and are required in many varieties (sizes and tread patterns). Using the modern supplier/customer partnership relations described in Chapter 12, flexible schedules can be developed for the supplier to ship specific tires at very short notice in quantities needed to meet current tractor assembly schedules. For such long-term arrangements, quantity discounts apply to total annual consumption of all tire models, rather than to individual orders. In both examples, inventory investment is significantly lower; the possibility of excess is reduced by gearing the *manufacture or procurement to definitive final assembly schedules.*

9 Making Outputs Useful

Garbage in; garbage out.

MRP: Tool of Many Uses

Early MRP programs were conceived and used as improved replacements for the earlier primitive and ineffective inventory planning and control practices discussed in Chapter 2. When first used, MRP was applied almost exclusively to order-release and due-date revision. As MRP was refined and as users gained experience, it became apparent that it yields information of value for several purposes other than inventory control. Also, users discovered that minor additional programs enabled MRP to serve as a planning system in areas well beyond the boundaries of traditional inventory control.

MRP programs properly designed, implemented, and used can function on three separate levels:

1. Planning and controlling inventories
2. Planning open-order priorities
3. Providing inputs to capacity requirements planning

MRP is a highly effective tool of manufacturing inventory management with the following benefits:

1. Inventory investment can be held to a minimum.

2. Planning is change-sensitive and reactive.
3. It provides future data on an item-by-item basis.
4. Inventory control is proactive, not reactive.
5. Order quantities are related to requirements.
6. It focuses on timing of requirements and order actions

Because of its focus on timing, MRP can do what none of its predecessors could, generate outputs that serve as valid inputs to other manufacturing logistics programs including purchasing, shop scheduling, dispatching and activity control, and capacity requirements planning. Sound MRP constitutes a solid basis for other computer applications in production and inventory control.

Production operations involve procurement of materials and their conversion into a shippable product. The principal outputs of inventory planning, whatever techniques are used, are purchase requisitions and shop orders, each calling for a specific quantity of some item by a date. Procurement and manufacturing can act only after inventory planning initiates such calls; it triggers activities and can be viewed as the upstream system.

Other planning activities along the two streams (procurement and manufacturing) are designed to extend and help execute the inventory plan. These downstream plans cannot improve or compensate for low-quality input they receive. Regardless of how well designed and implemented downstream planning programs may be, their effectiveness will depend on the validity, accuracy, completeness, and timeliness of their input from MRP.

MRP has the ability to suggest ordering, at the right time, the right items in the right quantities for delivery on the right date. It issues action calls according to a detailed time-phased plan that it develops. It keeps this plan up-to-date by reevaluating and revising it periodically in response to changes in the environment. It continuously monitors and adjusts all open-order due dates, reacting to such changes. With sound MRP providing inputs, downstream planning and execution programs can function effectively; without it they cannot.

MRP frequently is criticized for ignoring capacity constraints; it often plans production of items for which capacity is inadequate. Such criticism indicates lack of understanding of how it works, and its role. Programs can be designed to answer either the question of what can be produced with existing capacity or what should be produced to support MPS, but not both. MRP is designed to answer the latter question, showing what has to be done in order to execute the MPS plans.

It assumes that capacity constraints were considered when MPS were developed. MRP implicitly trusts MPS driving it; the validity of its outputs is determined by the MPS contents. Invalid MPS guarantee invalid MRP outputs. This is emphasized very strongly in Chapter 8.

Of course, invalid MRP outputs can be caused also by bad data in BoM, inventory, lead time, and activity-reporting files. The value of MRP depends on the quality of data it uses.

Uses of MRP Outputs

The primary outputs of MRP are the following:

1. Recommendations of planned order releases
2. Rescheduling notices changing open-order due dates
3. Notices to cancel or suspend open orders
4. Item-status-analysis backup data
5. Future planned order schedules

A variety of secondary or byproduct outputs can be generated by MRP at the user's option. It is not practical to list and describe all possible outputs and formats generated by MRP found in industry; MRP lends itself to tailoring, individualization, and infinite modification of outputs. MRP files (database), particularly inventory status records, contain a wealth of data, providing almost unlimited opportunities for the supplying of these data to a whole spectrum of possible outputs. Here are some of the common secondary outputs:

1. Exception notices, reporting errors, incongruities, and out-of-limits situations
2. Inventory-level projections (inventory forecasts)
3. Purchase commitment reports
4. Tracers to demand sources (pegged requirements reports)
5. Performance reports for people and functional groups

Six categories of outputs by functional use are

1. Inventory order action
2. Replanning order priority
3. Safeguarding priority integrity

4. Capacity requirements planning
5. Performance control
6. Reporting errors, incongruities, and out-of-limits situations

Inventory order action outputs occur when planned orders appear in current time-buckets. Other types of order action are increases, reductions, or cancellations of order quantities. The mechanics of obtaining these outputs are detailed in Chapter 4, and their use is discussed here and in Chapter 10.

Replanning order priority outputs alert inventory planners to cases of divergence between open-order due dates and dates of actual need, resulting from changed timing of net requirements. Examples of data triggering such outputs are presented later in this chapter. Standard MRP programs do not move released orders automatically, but outputs can indicate precisely how many periods each order affected should be rescheduled forward or backward so that planners can take the proper actions.

Excessive caution by planners can cause serious problems. Releasing an order to make 40 of an item instead of a planned 25 may cause problems with components, particularly manufactured items. Previously, based on the planned order for 25, a gross requirement for 25 sets of components was covered. However, an order for 40 will now be released with material requisitions for components to be filled by the stockroom. If any of these components has other parents with planned-order releases in the same period, its total gross requirements may now exceed its available quantity. The result is an unexpected shortage for which "that damned MRP system" is usually blamed.

For purchased items, the same problem may occur at suppliers' plants. Modern partnership relationships with suppliers include full communication of customer requirements and lean inventories in both parties' plants. Arbitrary changes such as that in this example will quickly destroy good working relations.

Outputs to help *safeguard priority integrity* aim at keeping order priorities honest, revealing inventory status problems caused by overstated MPS. Some companies use these reports to provide guidance for planners when accepting customer orders for guaranteed delivery. A "trial fit" of the order as an MPS item enables a net change MRP program to determine potential component shortages. If the order does not fit (too many shortages), the planner can recommend an alternative delivery date.

Capacity requirements planning outputs of MRP are open and planned shop orders in individual time periods, which are input into the load projection program as explained in Chapter 8. To keep load projections up-to-date and valid, they must be recomputed as MRP order schedules change.

Performance control outputs are comparisons of MRP plans with actual performance, enabling management to monitor the performance of inventory planners, buyers, the shop, suppliers, and cost accounting. The control-balance fields in net change MRP programs, discussed in Chapter 5, can be used to generate performance-control reports listing deviations from plan and their causes. Reports on item inactivity over a selected time, projections of inventory investment, and purchase commitments also belong in this category.

Inactivity (less than some preset low usage over a predetermined time) and obsolescence (no usage over a predetermined time) can easily be detected in time-phased MRP programs and reported for corrective actions, including disposal.

Inventory capital investment projections are easily made from MRP data giving projected on-hand quantities of each item multiplied by its standard cost, summarized for all items by time period over the planning horizon. More detail on this is given later in this chapter under "Budgeting Inventories" (p. 226).

Open and planned future purchase orders scheduled in time periods can be converted into purchase commitment reports for each supplier. The prices of such orders added in each time period for all suppliers represent the total cash the customer will need to pay in that period for these contracted deliveries.

Implosion (see Chapter 7) of a product's BoM will build up its cost from raw materials to end product using standard cost data in item master files. This reveals missing cost data and BoM errors that are obvious when similar components, subassemblies, or assemblies have significantly different costs.

Outputs *reporting errors, incongruities, and out-of-limits situations* are called *exception* or *action* reports. Some examples:

1. Date of gross requirement is beyond the planning horizon.
2. Number of digits of quantity in gross requirement exceeds size of the field.
3. Planned order is offset into a past period but placed in current period.
4. Number of digits of quantity of open order exceeds size of the field.
5. Number of digits of quantity of net requirements exceeds size of the field.
6. Number of digits of quantity of receipt overflows size of quantity-on-hand field.
7. Due date of open order is outside of planning horizon.
8. Allocated on-hand quantity exceeds current quantity on hand (potential shortage).

9. Past-due gross requirement has been included in the current period.

In addition to such exception reports, individual exception messages can be generated at the time inventory transactions are entered, listing reasons for transaction rejections. These and similar exception messages can be generated using the diagnostic routines and other checks discussed in Chapter 7. Some typical messages are

1. Part number is nonexistent.
2. Transaction code is nonexistent.
3. Part number is incorrect (using self-checking digit).
4. Actual receipt exceeds quantity of scheduled receipt by x percent (test of reasonableness).
5. Quantity of reported scrap exceeds quantity on hand.
6. Quantity of disbursement exceeds quantity on hand.
7. Order being released exceeds planned quantity.

Planning Valid Priorities

Traditional inventory control approaches were called "push systems" and "order launching systems" because they focused on getting orders started. They had to be supplemented by "pull systems," or expediting, to get orders completed at the time of actual need. MRP functions to both push and pull orders into and through operations based on start and need dates.

Sound priority planning and control of work in the factory depends on

1. Valid open-order due dates. These establish the relative priorities of orders, which must compete with other orders for productive capacity.
2. Valid order lead times. An individual order's actual lead time will vary with its priority; this is MRP's pull effect.

Over time, however, average actual lead times for an item must equal its average planned lead times. Chapter 11 discusses the corrective actions needed when they are not equal.

Each shop order involves a number of operations that must be performed to complete the order. A distinction must be drawn, therefore, between *order* priority and *operation* priority. Shop scheduling, loading, and dispatching techniques are based on operation priorities, valid only if derived from valid order priorities. MRP has the inherent ability to establish valid order priorities at the time of order release *if the MPS driving it is realistic and the BoM and inventory data it uses are accurate.*

MRP routinely reevaluates all open-order due dates for both purchase and shop orders in its netting process. It "knows" when an open order is not properly aligned with net requirements and can "tell" the user if programmed to produce this output report. This is covered in more detail with examples in Chapter 10 in the section "The Role of Inventory Planners" (p. 247).

MRP assists priority control by attempting to make two dates coincide: due dates and need dates. Due dates are those currently assigned to open orders, either put on the orders when released or revised by planners later. Need dates are MRP's latest indication of when orders actually are needed. These two dates often differ. MRP makes them coincide at the time of order release and reviews them afterward whenever a change in status causes a recomputation of net requirements. When MRP detects divergence of due and need dates it signals planners of the need to bring them back together by rescheduling actions.

When the dates diverge, need dates may move either forward or backward in time from due dates. MRP accordingly can signal either "expedite" or "deexpedite" the order. Expedited orders usually get attention; deexpedited orders usually are neglected. Both are important; it is obvious that priority control will improve if "cold" orders are moved out of the way of "hot" ones.

Valid priorities are impossible without accurate data. Significant errors in bills of material, inventory balances, open-order completion, and lead times will cause MRP to produce invalid priority plans. This topic is mentioned in several places in this book and discussed fully in Chapter 3.

Uses of Allocations

The MRP technique of allocating portions of on-hand inventory balances to "earmark" or "reserve" components for specific parent orders (see Chapter 5) has other important uses:

1. Verifying the accuracy of inventory records
2. Identifying present or potential shortages
3. Providing audit trails on issues from stores

Counting only a few carefully selected items each day, called *cycle counting,* the counters (storekeepers or specialists) need to compare their physical counts with record balances to detect errors and initiate a search

Assembly	Order	Qty	Part #	Qty short
B754	M5300	500	1975	500
			2244	1000
			5051	500
B780	M5000	700		
B826	M2100	200	1975	400

Figure 9-1. Using allocations for shortage lists.

for causes. As orders are released, requisitions are generated to withdraw components from stock. Because planners need to know true available balances, "withdrawals" are posted immediately to computer records as allocations and available inventory quantities are reduced. Some time may elapse, however, before components are withdrawn physically, during which record balances (available) and true stores (on-hand) quantities will be different. Allocations allow cycle counters to get record balances to compare to their cycle counts.

Before MRP programs were available, true shortage lists were based on physical counts of components made while staging (getting together) materials prior to filling stores requisitions. Modern MRP programs with accurate data in BoM and inventory records can be used during both planning and execution to show shortages, if any, on orders prior to release.

Allocating the quantity of each component needed to build a customer's order will reveal shortages such as those shown in Fig. 9-1 when the allocation produces a negative quantity available. The last column in the figure shows the gross shortage; part of this amount may be available, but planners, not computers, should decide which order should get this balance. Using these shortage data and deciding whether or not orders can be released in spite of them is discussed in Chapter 10.

Allocated balances also provide an audit trail to trace whether or not authorized issues have been made from the physical inventory. Balances remaining in allocation fields beyond a set time period are reported for checking.

Determining Capacity Requirements

MRP is capacity-insensitive; its function is materials priority planning, to determine how many of what raw materials and components will be needed and when, in order to execute MPS. MRP presumes that capacity limitations were considered in developing MPS and that they are realistic.

Long-range planning of capacity at MPS levels, called *resource requirements planning,* intended to ensure validity and realism in MPS, is covered in Chapter 8. Capacity requirements planning (CRP) is the function of determining required capacities of work centers, by period, in the near future to meet current production goals. This process uses MRP outputs of schedules for orders for component items to be produced, breaks these down into schedules of orders moving into and out of individual work centers, calculates hours of work (load), and accumulates loads by work center by period.

Fig. 9-2 shows a typical load report for a work center generated by this process. It is overloaded at the front, erratic, and drops off in future periods, because it contains only current open orders that will be completed soon. This pattern is still found in the load reports of many plants. Future planned but not yet released orders need to be considered also.

Planned orders were unavailable in pre-MRP practices and hence were neglected. Adding MRP planned orders to the capacity planning process produces a load report like the one seen in Fig. 9-3. It is even more front-end-loaded, still erratic, and shows an overload over the full term well beyond the work center's capacity. Most people viewing this report would conclude that the work center's capacity is inadequate—but they might be wrong!

Load calculations include only the real work of setup and running time. However, scheduling rules, used to determine the periods in which load is planned to fall, contain allowances for both move and queue

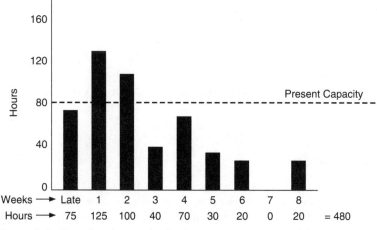

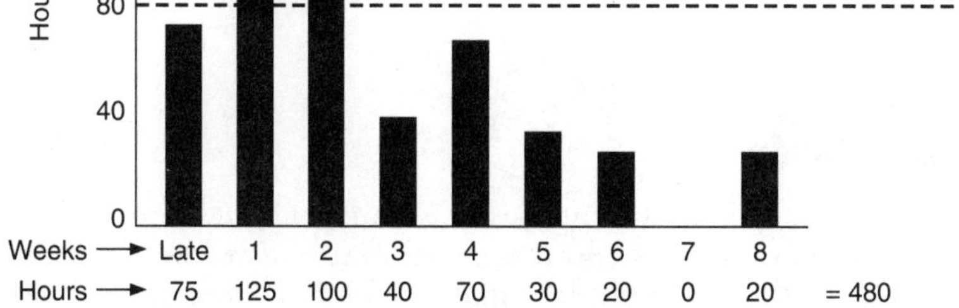

Figure 9-2. Typical work center load report.

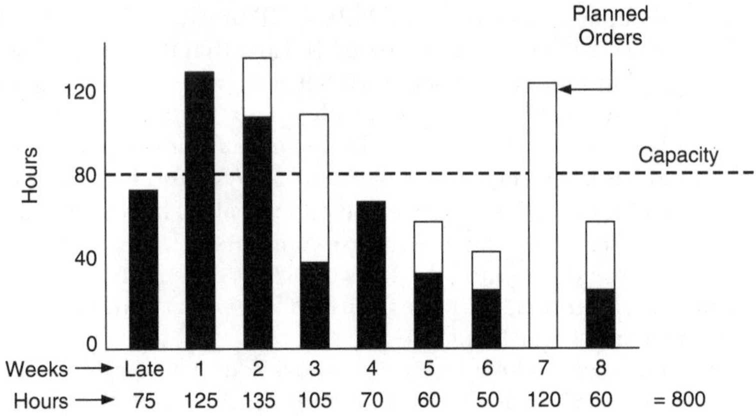

Figure 9-3. Load report including planned orders.

times. The latter recognize that load (work-in-process) will vary with the erratic flow of orders and that a planned cushion is needed to ensure that the work center doesn't run out of work.

The actual queue of orders at the work center accounts for the front-end load. Unless this exceeds the planned queue (say, 160 hours), no crisis actions are needed. The actual queue is included also in the total work for the 8-periods horizon. The planned queue (160 hours) must be deducted from the total load (800 hours) to determine how much of this total load is to be processed over the 8 periods. This 640 hours is the required throughput. The center's actual (demonstrated) capacity is 80 hours per period, and *its capacity is adequate.*

Problems of controlling capacity, other than not knowing how to determine capacity requirements from load reports, are discussed in Chapter 11.

MRP adds a dimension in short-range capacity requirements planning, providing data on future planned orders that complete the picture of loads on facilities. It also provides means for adjusting load data as requirements change and problems of production cause failures to execute plans.

Budgeting Inventories

Prior to MRP, budgeting inventories was independent of operating plans, being based on desired improvements in turnover rates, and was inflexible. Flexible budgets can be linked directly to operations using

MRP data, and can be revised to reflect significant changes. Projected on-hand balances in future time periods in component displays are expected inventory balances and are the best possible budget projection of storeroom inventories *if planning is sound and execution is effective.* Open and planned order data provide bases for calculating work-in-process inventory. Finished product inventory projections are shown in time-phased order point displays where this technique is used.

MRP Program Health Monitors

Outputs of MRP can be used to detect and help diagnose internal problems in operating programs. Problems can be grouped into four classes:

1. Data omissions and errors
2. Numbers of action notices
3. Transaction errors
4. Record errors

MRP programs commonly contain routines to report *omissions of essential data* in item master records. These data include lead times, lot-size rules, ABC class, costs, manufactured or purchased codes, order policy codes, and planner codes. Reports also can be generated showing manufactured parts with no BoM and purchased parts with BoM.

The *volume of action notices* provides signals of MRP health problems. Rising or excessive numbers of order-release or expedite notices may indicate overloaded MPS. Increasing numbers of delay notices may show the opposite problem, excessive capacity.

Transaction errors are routinely indicated by MRP programs. These include unplanned issues, reporting material moved from stores although MRP had not planned this, returns to stores caused by BoM errors or excess issues by stores personnel, and errors reported previously but not yet corrected.

Record errors are often revealed by MRP programs processing transactions. Negative balances in on-hand and open-order data are clear indicators of bad file data or poor handling of transactions. Quantities received on purchase or work orders in excess of ordered amounts show possible mistakes, sometimes in reporting the wrong unit of measure. Longstanding small balances in open purchase orders may be errors resulting from supplier deliveries on the low side of commercial tolerances.

Diligence in utilizing the capability of MRP programs to reveal actual and potential problems is at least as important as any other use of MRP outputs. Finding problems of data integrity early and fixing them fast will keep planning valid. As speed and flexibility become more essential to competing successfully, the importance of valid plans increases exponentially.

MRP is the primary element in the core system (see Chapter 1) for planning and controlling manufacturing logistics. It links parent and control items in properly time-phased schedules and enables plans to be revised as conditions change. It provides myriad data in integrated plans, which can be inputs to other planning and control activities as well as serve to monitor and measure performance of people and the business. No manufacturing company will be successful without a computer program that applies the logic of MRP to its logistics.

10
Planning versus Execution

A man plans his work, performs it, perfects it, and thus makes amends for the toiling and moiling.

ROBERT BROWNING

The Roles of Planning and Execution

Definitions of planning and execution were given in Chapter 1. Planning involves future activities and assigns numbers to quantify them. Execution concentrates on the present or the immediate future, attempting to convert plans into realities. Too often, users of MRP attempt to use this powerful program to do both. This is a costly mistake, reducing the effectiveness of both planning and execution.

The purposes of planning and execution also were stated in Chapter 1, emphasizing the dramatic difference. Planning is intended to *determine the amounts of resources needed to execute the plans.* Execution, however, *applies resources now available to produce what customers want* in the immediate future. MRP is an important element of the planning phase. It can assist in the execution phase by providing reliable, relevant data.

As was stated earlier in this book, plans will never be perfect; future events cannot be predicted exactly. It is equally certain that execution will encounter problems. Some planned resources will be lacking when needed for customers' orders. Planning must anticipate this and attempt to minimize the effects on execution. Execution must include actions to get deficient resources more quickly than planned or to use alternative ways to produce what customers want.

Confusing the roles of planning and execution and trying to use one MRP program to do both jobs has been widespread because of people's failure to understand their fundamental difference. A classic example is the application of "fences" defining a short portion of MPS as "firm," an intermediate section as "flexible," and the rest of the planning horizon as "unrestricted." The firm segment, equal to products' production cycle times, tightly limits changes to MPS because "it is too late to react to them." This is tantamount to saying, "If we haven't planned it, we can't make it," and "If we planned it, we must make it, even if we don't need it now." The idiocy of these statements is obvious to anyone familiar with real production.

The flexible segment covers future periods beyond the firm portion, in which there are limits to possible demands on resources that have little excess capacity and require long times to add more; expensive machines and test equipment are examples. In the unrestricted segment, almost anything is still possible.

The Planning Phase

Manufacturing planning, as discussed in Chapter 1, begins with Strategic, Business, and Production Plans. These are aggregate plans of market and business strategies, product families to implement strategies, and facilities to produce the products. These are top-management concerns and precede master scheduling and the detailed planning that follows.

Master production schedules are linked to detailed bills of material for production end items, not necessarily specific products to be sold to customers. Material requirements planning is the key element in all detailed planning, producing schedules for ordering materials, monitoring their progress, and revising schedules as situations change. Outputs from MRP initiate planning for resources other than materials: workers, machines, capital, tooling, and many more.

Planning attempts to provide answers to two basic questions:

1. Are we making enough in total?
2. Are we working on the right items now?

The first is the capacity question; the second relates to priorities. If the answer to either is "No," manufacturing will fail to achieve management's objectives. Integrated capacity and priority plans are needed to ensure that the answers are valid.

The Role of Master Production Schedules

Chapter 8 contains complete coverage of what MPS are and are not, how they are prepared and managed, and how management should use them. Together with their related bills of material, MPS are the most important set of data used in the planning phase. The following law governs the role of MPS:

> **Master production schedules are statements of what is planned to be and what can be produced, rather than what is now wished had been produced in the past or might be made now or in the immediate future.**

Those who fail to distinguish clearly between valid plans and wish lists, between planning and execution phases, their techniques and purposes, attempt to use the front end of the MPS in execution, substituting customer orders for planned items. Recognizing the serious problem of nervousness in MRP from MPS changes, they try to limit the planning horizon in which such changes can be made. This is futile. Customer orders drive final assembly schedules (see Chapter 8) and rarely belong in MPS.

The essence of the MPS law is that they must be realistic, a point made several times in this book because of its importance to sound planning and control. After over two decades of experience, this law is still honored mostly in the breach. There are three reasons: first, capacity requirements planning to test the realism of MPS is all too often neglected or poorly performed; second, few managers can resist the temptation to run a "lean plant" heavily loaded, so that no workers are idle and machine utilization is high; and third, no company ever refuses customer orders (and they shouldn't). Busy workers and machines are desirable goals, and the common measures of performance of production people. MPS are supposed to "exert pressure on them to get more product out the door." This approach to managing production has been called "brute force and ignorance." It is also counterproductive. There are much better ways with effective MRP programs to maximize plant output.

MPS can be used to aid in making promises (and revisions to them) to customers on future deliveries, and to alert managers that delays are affecting current plans. Where MPS end items, including option modules, are easily identified with shippable products, "Booked Orders" and "Uncommitted" lines as shown in Fig. 10-1 can be added to MPS displays. The latter, often called "Available-to-promise," projects the excess of available inventories above customers orders received. As shown, 15 Basic Machines remain unsold in week 5; however, 7 units of Option A are unsold in week 2, plus another 14 in week 5. It is time to make a decision on changing Option A's plan. Usually, possible completions of end items

Basic Machine

	Weeks										
	1	2	3	4	5	6	7	8	9	10	11
Master schedule	—	50	—	—	50	—	—	60	—	—	60
Booked orders	—	50	—	—	35	—	—	10	—	—	0
Uncommitted	—	—	—	—	15	—	—	50	—	—	60

Option A

	Weeks										
	1	2	3	4	5	6	7	8	9	10	11
Master schedule	—	35	—	—	35	—	—	40	—	—	40
Booked orders	—	28	—	—	21	—	—	5	—	—	0
Uncommitted	—	7	—	—	14	—	—	35	—	—	40

Figure 10-1. Using MPS to make customer delivery promises.

planned are doublechecked for shortages before promises are made to customers on new orders.

Closing the Planning/ Execution Loop

MRP converts MPS into detailed plans for execution and helps to monitor execution activities on component orders. The linkages between planning and execution are rigorous, and deviations can be made visible. Judgment should be exercised in sorting significant from trivial deviations in reports; this is done by "tolerance ranges" set by managers. Deviations within tolerance are not reported, since no action is required.

It is desirable and feasible to close the loop of these two activities at all levels. The status of execution can be fed back to planning so that actions to correct significant deviations can be taken promptly. Fig. 10-2 shows a typical high-level feedback report comparing actual production against planned amounts in MPS. The family totals on the last line can be compared also with Production Plans to alert people to capacity overloads. The Rollover column shows differences between planned and actual totals at the end of the preceding month.

Most production problems are caused either by invalid MPS or difficulties encountered in carrying out procurement and manufacturing tasks. Manufacturing logistics will be successful only if MPS are realistic in three ways:

		Rollover	Week 23	24	25	26	June	27	28
12 A 341	MPS		50	—	50	—	100	52	—
	Actual	—	48	—	52	—	100		
12 B 423	MPS		610	610	610	610	2440	610	610
	Actual	(10)	603	601	610	625	2439		
12 C 246	MPS		340	360	380	400	1480	420	440
	Actual	(12)	342	355	373	400	1470		
12 D 318	MPS		180	180	180	180	720	170	170
	Actual	5	185	176	178	180	719		
12 E 416	MPS		15	—	—	—	15	15	—
	Actual	—	—	15	—	—	15		
12 F 195	MPS		205	205	205	205	820	210	210
	Actual	8	206	202	203	207	818		
12 Family	Planned		1400	1355	1425	1395	5575	1477	1430
	Actual	(9)	1384	1349	1416	1412	5561		
Production plan							5600		

Figure 10-2. MPS versus actual production.

1. Needed materials are or can be made available.
2. There will be adequate time for operations required.
3. Capacity of resources is or will be available in time.

Each is equally important; lack of any one results in plans impossible to execute; then MRP and related planning programs become enemies. People responsible for production cannot be held accountable for executing invalid plans. Informal systems—side records, staging, hot lists, and expediting—replace impotent formal systems, and confusion, shortages, excess inventory, high costs, and dissatisfied customers result.

The last three serious situations are often believed to be separate problems; underqualified managers attack them individually with such edicts as "Cut inventory by 25 percent in the next 10 months!", "Reduce costs by 30 percent over the next year!", and "Improve customer service from 65 to 95 percent on-time deliveries in the next 6 months!" Surprisingly, each usually succeeds—but the other two objectives deteriorate! These three major objectives are interdependent and must be acted on simultaneously.

There are three classes of problems in the production process that can be attacked independently:

1. Inventory planning deficiencies
2. Falldowns in procurement
3. Execution failures

Inventory planning deficiencies show up as lack of coverage or insufficient lead time to cover net requirements. *Procurement falldowns* include past-due supplier deliveries, poor quality of materials, and inability to react to requested changes in order quantities or schedules. *Execution failures* are evidenced by late shop orders, unexpected scrap, or delays from lack of adequate tooling, machines, or other equipment capacity. Avoiding the first of these three problems is discussed later in this chapter; ways to attack the other two are presented in Chapters 11 and 12.

When a procurement falldown or a production failure occurs, information needed to take the proper corrective actions is not likely to be found in routine planning data. Very specific data, accurate and timely, are needed, providing answers to such questions as: What components are involved? What parent items are affected? After components are received, is there time to get parent orders back on plan? If not, how late will they be? What extra costs will be incurred? Who should be notified now? It is obvious that access to basic files—BoM, MRP profiles, open-order schedules, work center loads, cost data, etc.—can provide information, but skilled people must ask the right questions.

Keeping Master Production Schedules Valid

Problems in any of the three classes that cannot be solved at component levels should result in MPS revisions. Fig. 10-3 illustrates the logic and file relationships in tracking from low-level sources of problems to MPS. Chapter 3 shows how pegged requirements provide the means to trace from such components upward through BoM to the specific MPS needing change.

There are three types of capacity problems arising from overstated MPS: inadequate capacity over many periods; over a short range; or in specific work centers at a specific time. Figure 10-4 shows one MPS overstated in the first period; 6000 units are scheduled, compared with a six-month average of 3000. Not shown, but known to planners, is past actual output, which also has averaged 3000 units per period. Overstated MPS in current periods are quite common; many companies allow planned units not completed to accumulate in the first period, and the resulting loads often are equal to several periods' capacity.

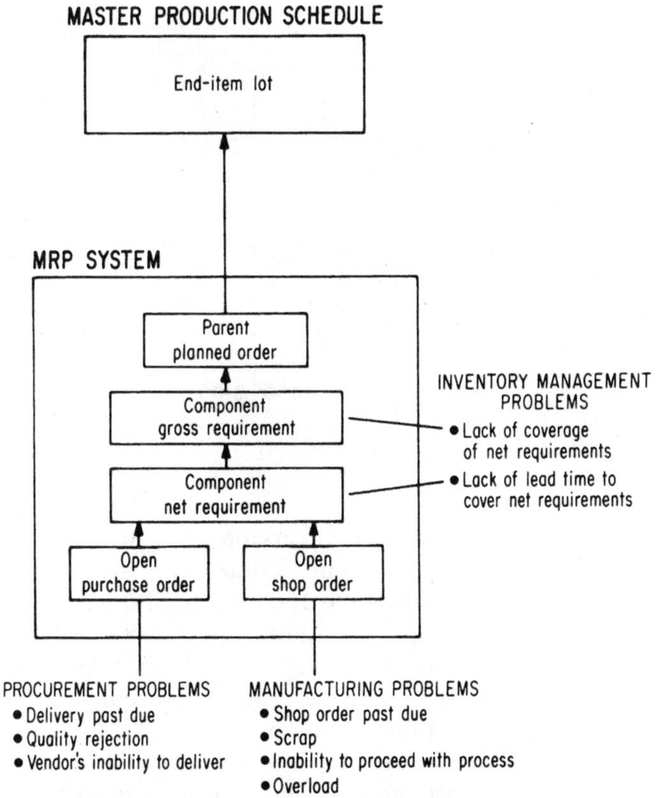

Figure 10-3. Relating production problems to the master production schedule.

		Period			
End item	6-months average production per period	1	2	3	4
A	100	180	100	90	100
B	200	480	200	160	180
Total	3000	6000	3000	3100	2900

Figure 10-4. A master production schedule overstated in the first period.

End item	6-months average production per period	Behind schedule	Period 1	2	3	
A	100	150	100	100	90	
B	200	180	240	200	160	
Total	3000	4500	2900	3000	3100	

Figure 10-5. A master production schedule overstated in the backlog.

The same problems arise when MRP users accumulate uncompleted units in a "past-due" or "behind-schedule" bucket, illustrated in Fig. 10-5. Here this backlog is 4500 units, equivalent to 150 percent of capacity per period; in addition, a full load is assigned to the first period. To get back on schedule the plant must produce items requiring 250 percent of its capacity in the current period! MPS must be realistic. Nothing can be produced yesterday, only today, tomorrow, or later.

Such significant overstatements of MPS have predictable results: most final assembly operations are past due, and most jobs in process are behind schedule and marked "rush"; expedite lists are long and are creating myriad crises, work-in-process inventory is excessive, and manufacturing costs are high. MRP output will be invalid and ignored; the formal priority planning system is useless at the very time it is needed most.

There are two common causes of overstated MPS: loading in too many MPS items, and having execution problems cause delays. Few companies have ever refused a customer order even though their plants are or would be overloaded. This is understandable, but making impossible delivery promises in the hope that somehow all orders will be completed on time is self-delusion. Customers will find out that their orders are late sooner or later; customer relations are usually improved when they know early the truth about deliveries. Good communications with customers eases and often eliminates overloads. Furthermore, removing overloads in planning always results in better execution.

Problems in execution often prevent making components, assemblies, and end products on schedule. Typical are shortages of material when suppliers fail to deliver, bad quality control with excess scrap or rework, poor tooling, machine or equipment breakdown, and absent workers. No planning technique can prevent such obstacles from developing or cope well when they do. Early MRP proponents lauded its ability to replan quickly after such upsets. Experienced users realized that this was no substitute for eliminating or at least minimizing the problems. This topic is covered in more detail in Chapter 11.

MRP provides a means for revising plans properly when such production problems cannot be eliminated and some item will not be completed as planned. Using pegging, planners can trace from the item upward through all BoM in which it appears. Sometimes solutions are found at an intermediate parent level, but if none is, some MPS must be revised and subsequently reexploded to establish up-to-date requirements and priorities on all components. It is not enough to reschedule only the order in question because of dependent priorities. This topic is discussed later in this chapter under "Priority Control" (p. 242).

When a component item will definitely be delayed, the true priorities of its co-components in their parent orders are now lower, since all are needed at the same time to make the parent. If MRP disregards this, and parents and MPS are not rescheduled, credibility of the formal schedule will be destroyed and it will be ignored. Failure to realign dependent priorities is the most common reason people find a formal priority system unreliable.

To keep the master production schedule in harmony with the realities of production is a classic problem of manufacturing management. With older, conventional methods it was difficult or impossible to identify the specific end-item lot (or several lots that use a common component) linked to some minor disaster on the shop floor or on the receiving dock. With material requirements planning, all the tools are there.

Revisions to MPS may result from marketing changes and also because of production problems. Customers change their orders, and marketing and sales people want special help from production to get a new customer order for some products. Such changes usually call for increasing the quantity or advancing the timing of an end-item lot. These are desirable changes, but they cannot be made arbitrarily without revising MPS.

Manufacturing companies fighting to succeed (in some cases to survive) in highly competitive global markets must be flexible in reacting to marketplace changes. Computer-based MRP and other planning programs can be revised at blinding speeds. Plants and their suppliers are not so flexible, however; their execution of plans is constrained by the availability of resources. Each schedule change requires an analysis of its effects on resources; this is part of replanning and can be done quickly. The limits of flexibility in execution narrow with passage of time, making it less and less practical to effect changes as items near their scheduled completion dates.

Obviously, changes in MPS beyond the total cycle time of any product (the "flexible portion of MPS" discussed earlier in this chapter) can be allowed, within reasonable limits, because few, if any, real commitments of resources have been made and there are no immediate effects on open orders. Replanning is cheap, quick, and easy. Within these cycle times (the "firm portion"), however, execution will have begun, resources will have been committed, or additional ones will be required to support MPS changes. Inability to get added resources may preclude making the changes. High costs of meeting revised schedules may exceed the potential benefits. Altering execution activities can be expensive, slow, and very difficult.

The recognized primary strategy in production now is:

Don't commit flexible resources to specific items until the last possible moment.

Resources are flexible until committed to some item. Raw materials can be made into many components, common parts into many subassemblies, and these into many products. People can perform many different operations. Machines can produce wide varieties of components. Once execution begins and resources are used to buy or make specific items, some flexibility is lost. After unsold finished products are in stock, the only flexibility remaining is to find some "least worst" way to dispose of the excess. Having excess inventory, called *safety stock,* is the most expensive and least effective cushion against market changes.

The Execution Phase

Recently, attention has shifted from getting more powerful and sophisticated planning programs to speeding up and smoothing out flows of materials and data. This resulted from recognition that shorter cycle and lead times were more beneficial than attempting to cope with chaos by carrying safety stocks and replanning often. Attention shifted from planning to execution.

Excessive zeal in promoting the priority planning power of MRP programs, coupled with lack of understanding of the need for sound capacity planning and control as well as tight priority control, led many people at all organization levels in manufacturing firms to attempt to use MRP in both planning and execution phases. This confusion was intensified by manufacturing resources planning (good name) being called MRPII (confusing acronym).

Many underqualified American managers lacked full understanding of how manufacturing works; they hoped to buy their way out of trouble with MRP and MRPII programs. The search for and trial of these and other quick fixes devoured money, time, and other resources, with a few outstanding successes, a few disasters, and a large number of partial failures.

As an example of misuse of MRP as an execution tool, marketing-motivated changes were introduced into MPS to "see how they fit." MRP was then run as a simulation (no problem; it really simulates in normal runs), to indicate the effects on open orders. Analyses of these effects, however, were neither simple nor conclusive. Neither was the task of restoring the initial conditions if the simulation proved unacceptable. Actually, the approach was unnecessary; better ways are available using execution techniques rather than replanning.

In manufacturing businesses where products sold to customers or their major subassemblies comprise MPS, a simple match can be made of incoming customer orders or customer-oriented changes with MPS items. This trial fit indicates which orders or changes may be accepted with the customer's requested delivery date and which need further investigation. Sometimes one or another customer will accept a later date, making the MPS-fit good, and the problem is solved at the top planning level. This assumes, of course, that the plan will be executed well and fully.

Another way to determine whether or not customer orders can be built and shipped on time *even if they have not been planned in MPS* is to test for shortages using the allocation method described in Chapter 9 and illustrated in Fig. 10-6. Today's date is February 1st. It appears that

Assembly	Order	Qty.	Part #	Qty. short
B754	M5300	500	1975	500
			2244	1000
			5051	500
B780	M5000	700		
B826	M2100	200	1975	400

Part #	On hand	On order Qty. date	Assembly	Date due	Qty. avail	Qty. short
1975	1200	1500 3/5	R530	2/1	700	
			R450	2/8	400	
			B754	2/9		500
			B826	3/2		400

Figure 10-6. Using allocations to schedule assemblies.

Order M5000 for assembly B780 is the only one that can be built; it has no shortages.

An experienced planner would look further. Part 1975 is a shortage on both M5300 and M2100, and additional data on it, all available in MRP programs, are shown in the lower part of the figure. There are 1200 units on hand and an additional 1500 units are on a work order due to be completed on March 5th, about four weeks hence. Other assemblies already have claims (allocations) on Part 1975; assembly R530 needs 700 immediately, R450 will need 400 next week, using up 1100 of the 1200 total on hand. M5300 (one of the orders being studied) also will need 500 next week, and M2100 (the second order studied) will need 400 in four weeks. What must be done to hold all three order schedules?

1. B826 can be built on March 2nd if the open order for 1500 units of 1975 is scheduled for completion then (hard execution data) instead of March 5th (soft planning data). This should be easy, and there are no other shortages on B826 order.

2. B754 can be built on February 9th if 400 units (its gross shortage is 500, but 100 will be available after R530 and R450 have been built) can be split off the 1500 unit order and completed by then. Whether or not this can be done can be investigated using current order status and load data on operations and work centers involved. The other two shortages, Parts 2244 and 5051, can be checked the same way for actions needed to get them on time.

MPS and MRP plan production activities, but other data drive execution. Customers' orders like those in the preceding example are prime drivers; MRP allocation programs serve as excellent supporting tools. Other execution drivers are MRP recommendations to release planned orders, shop-tracking early- and late-order signals, and final assembly schedules. Signals of delays or interruptions in order flow should trigger prompt corrective actions. Tight execution depends on accurate, timely (hard) data on order status, correct order priorities, and fast action.

Input/Output Capacity Control

Correct order priorities, however, cannot be maintained if capacity to produce *everything planned* is not available when needed. Techniques for planning capacity, both rough-cut and detailed, are covered in Chapter 8, which includes both their mechanics and applications. Capacity control has two phases: input and output. Both must be well planned and tightly controlled if operations are to be effective.

Figure 10-7 is a typical input/output control report for a starting (gateway) work center, showing planned and actual data and cumulative

(Work center 3
All data in standard hours)

	Week						
	9	10	11	12	13	14	15
Input							
Planned	270	270	270	270	270	270	270
Actual	275	265	230	255			
Cumul. Dev.	+5	0	–40	–55			
Output							
Planned	300	300	300	300	270	270	270
Actual	305	260	280	295			
Cumul. Dev.	+5	–35	–55	–60			
Queue							
Planned (150)	240	210	180	150	150	150	150
Actual (270)	240	245	195	155			

Figure 10-7. A typical input/output control report.

deviations for both input and output plus planned and actual queue. A decision has been made to cut the queue in half over the next four weeks. The required throughput rate to support the MPS is 270 hours per week, but output is planned at a rate 30 hours per week higher (300 hours) to work off the queue.

Control of input involves selecting orders with the right priority and releasing them at nearly, not exactly, the right rate. Figure 10-8 shows MRP recommendations for the next two weeks. The orders recommended for release in week 9 total only 198 hours; the planner then looks at week-10 orders for one (Part 3641) to make up the planned 270 hours release rate. The total of 275 hours released is posted in week-9's actual input bucket.

In week 9, the work center completes 305 hours against the planned rate of 300; this is posted to its actual output bucket. The actual queue was cut to 240 hours as planned. The next three MRP week-10 recommended orders (4130, 2317, and 4537) total 265 hours (close enough to the planned 270) and are released and posted to the I/O report; the remaining week-10 order for 6853 is held to avoid inflating the queue.

In week 10 the work center has problems and completes only 260 hours, falling 35 hours behind plan, and the queue increases. The next step is crucial. Releasing the planned 270 hours will compound the excess queue problem and probably result in poor priority control, since the work center has more chance to select the wrong orders to work on. The right rule to follow is

Input must be equal to or less than output, never more!

Work Center 3 Week 8				
Planned rate = 270 hr				
Current backlog = 270 hr				
		Order quantity		
Part #	Rels. week	Pcs	Hrs	Cumulative hours
5252	9	450	37	37
1824	9	610	113	150
7266	9	315	48	198
3641	10	580	77	77
4130	10	220	53	130
2317	10	1100	127	257
4537	10	725	85	342
6853	10	500	23	365
5348	11	150	62	62

Figure 10-8. MRP order recommendations.

The I/O report shows that this planner followed that rule for weeks 11 and 12, with the result that the queue is only 5 hours above plan! Very good. Total input, however, is 55 hours less than planned, and work center output is 60 hours short. The planned rates of input and output *to support the MPS* were 270 hours per week; the work center achieved more than this output, but it must make up the 60 hours deficiency against plan soon.

The input deficiency is almost unimportant; pushing out orders doesn't increase output, it just builds work-in-process queues and gives the priority choice to the wrong people. The shorter queue after week 12 and the planners' choice of orders to be released lessens the opportunity for work on the wrong ones.

This example illustrates how input/output data are used to monitor throughput rates and trigger fast response to problems so that priorities of work are more valid. The example used a starting work center; the process for intermediate centers is identical, except that control of the mix of input orders is difficult. Actual hours input and their effect on queues can be posted to I/O reports, and will signal problems there very quickly.

Priority Control

Manufacturing companies may be divided into four categories, based on their products and stocking policies, as follows:

1. One-piece products made to order
2. One-piece products made to stock
3. Assembled products made to order
4. Assembled products made to stock

Companies of the first type (e.g., foundries, crankshaft manufacturers) have the simplest order-priority problem. Customer-requested order delivery dates, when confirmed, represent the priority. Unless customers change order due dates, priorities remain fixed. Relative priorities of all open customer orders are established by their respective delivery dates.

Firms of the second type (e.g., fasteners, hardware) have a little more complex order-priority problem. To keep these priorities honest, open-order due dates must be related to the availability of stock for each item. Priorities are vertically dependent on the customer demands that cause stock depletion.

In the third type (e.g., machine tools, large home appliances), the priority problem is quite complex. Orders for components of assembled products have horizontally dependent priorities. The availability of all components is a prerequisite to the completion of each parent subassembly and of the end product.

The priority problem is most severe in the fourth type (e.g., hand tools, power saws), because numerous common parts make order priorities both vertically and horizontally dependent. The true priority of a component shop order is a function of the available supply of the end product and the availability of all other components required for assembly of its parent items.

This classification is oversimplified; many manufacturing businesses do not fit neatly into any one of the categories. Nevertheless, it is useful for analysis and exposition. Obviously, the more severe the priority problem, the greater the need for effective MRP replanning capability.

The true priority of orders (when they really are needed) may differ from the planned priority assigned by MRP. The former depends on availability of all components at the time the parent orders should start into production; the latter is determined by MRP based on parent item order planned start dates and *an assumption that all components will be available at that time.*

For example, inventory of product A, which is shipped from stock, is forecast to run out in 6 weeks, and a replenishment order has been planned for release in 4 weeks. An order for a component has been started with a due date 4 weeks hence. If sales lag and product A is still in ample supply as this component order nears completion, its true priority is lower than its due date indicates. This is *vertical priority dependence*; true priority is a function of availability of items on higher levels in the product structure.

Components of assembled products have *both vertical and horizontal priority dependence.* Some components are not really needed if all components will not be available for the scheduled assembly of their parent item. For example, Fig. 10-9 shows three manufactured components, A, B, and C, of parent X. Orders for the three have different lead times but identical due dates, which coincide with the scheduled start of assembly of parent X. If the order for component A, for instance, is unavoidably delayed (possibly scrapped) and has too little time for on-schedule recovery, parent X will not be assembled on the date planned. The true priority of orders for B and C has therefore dropped; they will not be needed on the original order due date.

MRP is oblivious of this fact if the indicated requirements for B and C have not changed. MRP users are responsible for reestablishing priority integrity by rescheduling parent X's planned order (using the firm planned order technique described in Chapter 3) and having MRP replan requirements and need dates for all component orders. Rescheduling X order may delay its parent and others up to the MPS, and it may even have to be changed to maintain priority integrity.

Planners know immediately that scrapping a component order will delay assembly of its parent, but they also need to know

1. Which parent item?
2. How many component orders are affected?
3. What are the identities of affected orders?
4. How should each of these orders be rescheduled?

Well-run MRP provides answers to these questions quickly and accu-

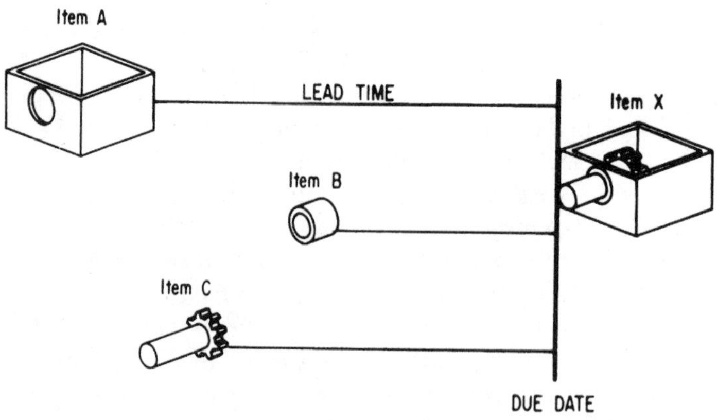

Figure 10-9. Horizontal priority dependence.

rately when MPS are made realistic; it recomputes requirements, suggests rescheduling affected open orders, and shows for how many periods each individual order should be rescheduled.

The example in Fig. 10-9 is simplistic. The complexity of even this problem, however, is compounded: component *A* may have multiple parents, some not affected by the scrapped order; several other orders for *A* in different stages of completion may be open; and lot sizing may obscure the proper rescheduling action. Real products will involve hundreds of items and dozens of open orders and pose massive problems of analysis.

This formidable rescheduling problem seems to defy solution, particularly since scrapping orders, equipment breakdown, late supplier deliveries, and other production problems occur frequently, delay completion of orders, and destroy priority integrity. MRP is capable of helping planners solve such problems, but few ever learn how or take the time to use it for this. Invalid order priorities are a classic, chronic, and intractable problem in industry, even with computer-based, time-phased MRP. Four conclusions are evident:

1. The process of maintaining priority integrity can be extremely tedious and time-consuming.
2. The effects of upsets can be very widespread.
3. Working energetically to get *A* back on schedule will yield better results than giving up on it and rescheduling.
4. Eliminating common production problems is the best solution of all.

MRP, when used properly, provides excellent priority plans, but these must be supplemented by effective priority control in the factory. MRP can plan order priorities correctly and keep them up-to-date by replanning order due dates. Obviously it cannot ensure that these due dates actually are met. *Priority control must provide the procedural machinery to execute plans.* This is variously known as dispatching, job-assignment, shop floor control, and production activity control. Four functions are involved:

1. *Advance planning*—providing information on orders due to be released or to arrive soon in work centers, early enough to enable people to get necessary resources ready
2. *Order selection and assignment*—indicating a desirable sequence of running orders, and providing information to aid in assigning orders to particular machines or individuals
3. *Feedback*—showing results of execution activities in locations, piece counts, and costs of processing orders
4. *Priority updating*—keeping order priorities valid

Priority control devices exist in some form in every manufacturing plant. In pre-MRP times, dates on order packets that authorized work and described the sequence of operations served to set priorities. Without input from MRP, however, they were static, becoming quickly out-of-date and invalid. Shop personnel then developed informal priorities based on shortage lists.

Chapter 9 defines two types of priorities: order and operation. The former are determined by order due dates set by MRP when orders are released, or by need dates from MRP after replanning. Scheduling rules use estimates or standards for each element of lead time (setup, run, move, and queue times) to convert order due dates into operation scheduled start and completion dates. In addition to assigning loads to time periods in capacity planning, these dates determine operation priorities.

The modern instrument of priority control is a dispatch list or departmental schedule, typically generated daily or weekly. The latter frequency is less effective, since tight priority control requires more precise data. It may be a printed report, cards, individual work-sequence messages printed via a remote terminal, or a listing on a CRT screen. A sample dispatch list is shown in Fig. 10-10. Its essence is a ranking, by relative priority, of work orders physically present in a work center. When

DISPATCH LIST

Department: **No. 12, Turret lathes** — Date: **447**

Order #	Part #	Quantity	Operation #	Start date	Std. hours	Remarks
Jobs in department						
5987	B-3344	50	30	441	3.2	*Tooling in repair*
5968	B-4567	100	25	444	6.6	Eng. hold
5988	F-8978	30	42	446	4.8	
5696	12-133	300	20	447	14.5	
5866	A-4675	60	20	447	7.0	
5996	A-9845	200	30	448	6.2	
5876	F-4089	25	40	480	5.4	
Jobs scheduled to arrive this date						
6078	A-3855	160	30	446	6.5	In dept. 15
6001	D-8000	300	65	448	9.8	In dept. 08

Figure 10-10. Daily dispatch list.

a significant amount of preparation is necessary to begin work on orders, it also includes released orders soon to arrive in the work center. Planned orders rarely are included.

The sequence of orders on dispatch lists is based on operation priorities, which in turn are derived from MRP order priorities. As order due dates are changed by MRP, operation priorities must be revised accordingly. They may be based on operation start dates, as in Fig. 10-10, operation finish dates, or on calculated ratios, some of which relate work content of orders.

Shop performance on these orders is tracked from daily production reports generated by a wide variety of data collection equipment, from simple handwritten lists to bar codes on dispatch lists read by light pencils. Actual output is compared to planned, and significant deviations are reported for corrective actions. These are factored into revisions to daily dispatch information.

Purchase orders, unlike work orders, do not involve operation priorities but require similar order priority data. Purchased components have dependent priorities, the same as manufactured items, which can be planned properly by MRP.

The Role of Inventory Planners

The first steps in execution activities include the following:

1. Release orders for production.
2. Send requisitions to purchasing.
3. Change order and requisition quantities, or cancel them.
4. Reschedule open shop orders.
5. Request changes in open purchase order timing.
6. Approve requests for unplanned stock disbursements.

People performing these tasks are known as *inventory planners, inventory analysts,* and *inventory controllers,* and they are responsible for planning and controlling a specific group of items. In addition to the listed execution activities, they also perform tasks related to the inventory files; they

1. Handle engineering changes in bills of material of items under their control
2. Monitor inventory for inactivity or obsolescence, and recommend disposition
3. Investigate and correct errors in inventory records

4. Participate in analyses of cycle inventory counts
5. Analyze discrepancies in item requirements and coverage, and take appropriate corrective actions
6. Request changes in master production schedules

All of these duties are essentially the same with MRP as they were before MRP. When this powerful program became available, it was seen as the tool of this group only. They interacted continuously with MRP, received its principal outputs, and took action based on data it supplied. They extracted from its files data needed for analyses and supplied data to keep it current.

Most of these tasks are routine and self-explanatory, but a few warrant a more detailed look. Transactions continually modify inventory status data, which in turn provide clues to action. The principal actions signaled by MRP "action notices" are these:

1. Release planned orders.
2. Expedite released orders.
3. Delay released orders.
4. Cancel released orders.

These involve the timing of orders; changing quantities once work has begun on orders is limited to splitting a lot into two or more sublots with different schedules. It is usually difficult and costly, and may even be impossible, to increase the quantity of an open purchase or work order. It is undesirable to split work orders because the additional setups aggravate capacity problems, which often are the causes of people's desires to split orders.

In practice, order-related actions generally are limited to

1. Releasing orders in the right quantity at the right time
2. Rescheduling due dates of open orders as required, to make them coincide with the dates of actual need

In both, inventory planners can lean on MRP, which suggests both the quantity and the timing of planned-order releases and which also constantly monitors the validity of all open-order due dates. The following two sections illustrate how MRP determines when to signal "Release the order" and "Reschedule the order."

Releasing Planned Orders

MRP programs consider planned orders to be ready for release when a quantity appears in the current-period bucket. This happens either as a result of offsetting for lead time in the requirements explosion or by passage of time gradually bringing a planned-order release toward the current period, called the *action bucket.* MRP tests the contents of this bucket; when a quantity appears, as shown in Fig. 10-11, it generates a message to inventory planners recommending "Release the order."

Inventory planners review these recommendations and decide whether or not to execute them. They may override MRP, holding orders if starting work centers cannot handle them now, or releasing orders early to level-load work centers as illustrated in the input/output control example earlier in this chapter. They may also change quantities if amounts of available raw material or components are not enough to make the full quantity planned, reducing the quantity rather than delaying order release if this is preferable. Conversely, planners may increase order quantities if a parent needs more. Experienced planners know that this may cause a shortage of components needed to make other parent items. Releasing planned purchase orders is simpler for two reasons: first, no component materials are involved, and second, planners are less concerned with level-loading suppliers' facilities.

In net change MRP, covered in Chapter 5, changing planned-order quantities at the time of shop-order release upsets the interlevel equilibrium that the program strives continuously to maintain, introducing errors into the records. Experienced planners immediately correct these using special transactions.

When the planned order shown in Fig. 10-11 is released, the transaction changes the status of the item as shown in Fig. 10-12. Both open orders are now scheduled with the correct due dates. Changes in gross requirements, as shown in Fig. 10-13, will change the due dates in this example but only the timing, not the quantity. Coverage is still

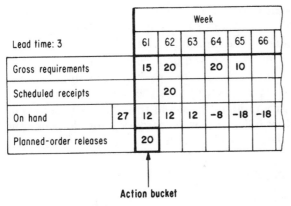

		Week					
Lead time: 3		61	62	63	64	65	66
Gross requirements		15	20		20	10	
Scheduled receipts			20				
On hand	27	12	12	12	-8	-18	-18
Planned-order releases		20					

↑ Action bucket

Figure 10-11. A planned order mature for release.

+20

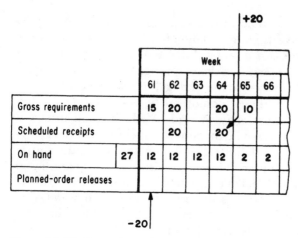

		Week					
		61	62	63	64	65	66
Gross requirements		15	20		20	10	
Scheduled receipts			20		20		
On hand	27	12	12	12	12	2	2
Planned-order releases							

−20

Figure 10-12. Status change following planned-order release.

		Week					
		61	62	63	64	65	66
Gross requirements		30	5		10	10	10
Scheduled receipts			←20		20→		
On hand	27	−3	12	12	22	12	2
Planned-order releases							

Figure 10-13. Open orders do not coincide with dates of need.

adequate—no additional orders are required—but one open order is now too late and the other too early. MRP will signal the need for rescheduling.

An additional powerful action notice is rarely used: "Prepare to release." A significant element of lead time, listed with others in Chapter 8, is preproduction time. This is time required to prepare paperwork (purchase and stores requisitions, shop order packets, etc.) and to check availability of tooling, test equipment, process specifications, and other resources. MRP can be programmed to signal the time to begin such actions using its forward visibility of planned orders. This element can then be eliminated from planned lead times, making them much shorter.

Rescheduling Open Orders

Net change MRP detects the need to reschedule open orders immediately upon processing the transaction that caused gross requirements to change;

regenerative MRP detects this during the requirements planning run. Changed gross requirements necessitate recomputation of projected on-hand data, and these give clues to the action required by planners. Figure 10-13 shows a net requirement of 3 in the first period, with two open orders for 20 in weeks 62 and 64. When open orders can cover net requirements if rescheduled, MRP will not generate a new planned order but will recommend rescheduling of the nearest open order.

Week 63 projects 12 on hand, and the gross requirement in Week 64 is only 10; this can be covered in full by the quantity on hand and there is clearly no need now for the open order to arrive in week 64. This order should be rescheduled to week 65 when it is actually needed. The two tests for open-order misalignment are

1. Is there an open order scheduled in a period later than a net requirement?
2. Is there an open order scheduled in a period with no net requirement (the gross requirement is equal to or less than the actual on-hand quantity at the end of the preceding period)?

MRP carries out these two simple tests whenever netting is recomputed. If a test is positive, the system generates the appropriate rescheduling message. The extension of the second test to subsequent periods will indicate that an open order should be canceled when the projected on-hand quantity in the period preceding the scheduled receipt of the order is sufficient to cover all remaining gross requirements. MRP automatically recomputes planned-order schedules to align them properly with net requirements, and no inventory planner actions are required.

To maintain true relative priorities of both open shop and purchase orders, planners must reschedule due dates for orders now needed earlier or later. Prompt reactions to increases in requirements and signals of earlier order completion are very desirable to prevent shortages. Pressures of planner work load, combined with lesser concerns about high inventories, result in neglecting orders arriving too soon. This in turn often results in plant overloads characterized as "trying to get six pounds out of a five-pound sack" when the extra pound is not really needed.

Changes in inventory status and schedule performance occur every minute of the day; in well-managed companies most are planned, and they trigger no action notices. Transactions representing unscheduled or excessive stock disbursements, unexpected scrap, and physical inventory adjustments can largely be minimized and even eliminated, ending the MRP signals they generate. When production operations "get their house in order," most of the causes of nervousness go away.

Inventory planners notify MRP of revised shop-order due dates, it adjusts its data, and the revised dates are used in operations scheduling and dispatching programs. Remaining operations on these orders are rescheduled properly, and the new operation dates (start or finish, depending on users' choice) are used in dispatching and also to recompute work load. Ratios setting priorities for dispatch lists (rather than only operation start or finish dates) are calculated using the new order due dates. Inventory planners notify purchasing people of revised need dates; only when the latter act to get supplier acceptance does MRP reflect the changes.

Planners may decide not to advance order due dates (contrary to MRP recommendations) when there is safety stock, or when the earlier date would be impossible to meet. In the latter case the proper course of action (whether practical or not is another question) is to peg upward in an effort to solve the problem, possibly all the way to MPS, which may have to be changed.

Covering Net Requirements

Probably the most serious problems inventory planners must cope with are discrepancies or misalignments between net requirements and orders covering them. These result from unplanned events changing gross requirements. Planner actions are limited; they cannot change actual quantities of gross requirements directly. They can change only the timing and/or quantities on orders. To change gross requirements for components, they must change the planned-order schedules of their parents.

Two features of modern MRP programs described in Chapter 3—pegged requirements and firm planned orders—help them to make such changes correctly and effectively. How these are used is illustrated in three figures. Figure 10-14 shows status data for raw material *Y*, a purchased material with five weeks' lead time. Figure 10-15 shows that gross requirements for *Y* in week 31 increased from 20 to 30 because of a planned order increase in its parent, fabricated part *X*, and MRP recommends an immediate order release for 35 units. The inventory planner's review of this recommendation prior to releasing the order reveals a problem: *Y* has a five-week procurement lead time but is now needed in three.

Can a "rush" purchase requisition be avoided? The planner pegs to *Y*'s parent *X* to find the source of its gross requirement in week 31, sees that the 30-unit planned order for *X* there covers net requirements of 30 in weeks 33, 34, and 35. The rush purchase order can be avoided if the planner reduces the first planned order for *X* from 30 to 20, designating it as a firm planned order, thus preventing MRP from increasing it again

Lead time: 5

		Week							
		28	29	30	31	32	33	34	35
Gross requirements					20			25	
Scheduled receipts				15					
On hand	5	5	5	20	0	0	0	-25	
Planned-order releases			25						

Item Y

Figure 10-14. Original status of item Y.

Lead time: 2
Order quantity: LTC

		Week							
		28	29	30	31	32	33	34	35
Gross requirements		3	7	10	6	8	10	14	10
Scheduled receipts									
On hand	38	35	28	18	12	4	-6	-20	-30
Planned-order releases					30			25	

Fabricated part X

Lead time: 5

		28	29	30	31	32	33	34	35
Gross requirements					30			25	
Scheduled receipts				15					
On hand	5	5	5	20	-10	-10	-10	-35	-35
Planned-order releases		35							

Raw material Y

Figure 10-15. A problem of coverage.

to 30 during the next replanning cycle. MRP then advances by one week the second planned order for 25 units of *X* to cover its gross requirements. *Y*'s lead time can accept this. Figure 10-16 shows the two records after these changes.

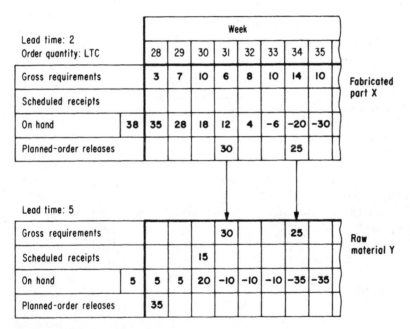

The same type of problem would have arisen if the supplier of *Y* had informed the planner that he was unable to ship the order for 15 units in week 30. If *Y* were a fabricated part, scrapping 10 of the 15 in process would have the same effect. In this example, the inventory planner was

Fabricated part X

		Week 28	29	30	31	32	33	34	35
Gross requirements		3	7	10	6	8	10	14	10
Scheduled receipts									
On hand	38	35	28	18	12	4	-6	-20	-30
Planned-order releases					(20)		25←		

FIRM PLANNED ORDER (circled 20)

Raw material Y

		Week 28	29	30	31	32	33	34	35
Gross requirements					20		25		
Scheduled receipts				15					
On hand	5	5	5	20	0	0	-25	-25	-25
Planned-order releases		25←							

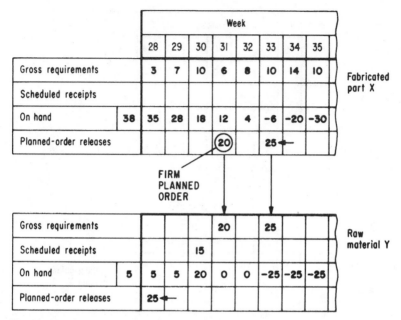

Figure 10-16. Solution of problem of coverage.

able to reduce the parent planned order quantity because lot sizing covered net requirements of several periods. A scrap allowance or safety stock would also provide an excess over net requirements. These alternatives available to planners show clearly why computer programs should not be allowed to make arbitrary changes without manual review.

Such problems can be solved, or at least eased, by changing the timing, not the quantity, of parent planned orders if the lead time can be compressed (See Chapter 11). Planned-order releases can be made with less than planned lead time, using the firm planned-order feature of MRP to hold the correct need date. Reducing the lead time of the planned order for *X* in Fig. 10-16 by one week, for example, moved *Y*'s net requirement back one week, allowing an extra week for material procurement.

Many transactions may affect the same inventory record on the same day; components common to several parents are an example. Such transactions in net change MRP may revise the timing of open orders several times a day, with some changes canceling others. Planners' actions, however, can be decoupled from the rate at which individual changes occur and are processed by the MRP program. The most common method of doing this in practice is programming MRP to accumulate action requests and present them to planners periodically.

The period used will depend on the type of action. Weekly reports of orders to be released are usually adequate except when production cycle times are very short, as in food processing and electronic assembly. Revised due dates for open shop orders may be reported once per shift, to maintain validity of shop priorities. Some action messages, however, should be generated without delay because the need for corrective action is critical. For example, an open purchase order may become a candidate for cancellation as a result of lowered requirements. A 24-hour delay in notifying the supplier can incur significantly higher cancellation charges and may even make the difference between being able to cancel or not. Similarly, reaction without delay to allow maximum recovery time is vital when excessive scrap occurs, or when there is a significant loss of on-hand inventory following a physical count.

Coping with Nervousness

When net change MRP programs process major changes in MPS, thousands of records can be affected and the status of an inventory item may change several times. This is an unusual situation, and extreme nervousness can be avoided by suppressing all action notices until processing has been completed. Among the many causes of "nervousness" in MRP are changes in MPS, supplier deliveries, material quality, order quantities, safety stock, lead time, and design, together with record errors and unplanned transactions.

Nervousness in normal operation poses difficult choices both for regeneration and net change MRP users. Do they prefer using data not quite up-to-date, or do they want to ignore action signals from MRP? Ideally, data should be current and people should be able to react to all signals; both rarely occur. Usually, the two halves of the question are addressed separately.

Most MRP users start with regeneration because it is easier to understand and implement than net change, is more forgiving of lapses in discipline in data handling, and emits action notices only once a week, not continuously. Net change is adopted only after considerable MRP experience leads to a conviction that more up-to-date planning is needed. Few can handle the flood of signals emitted by it, and most have to make compromises. Usually they conclude that deliberate avoidance of some user actions, with full knowledge of current facts, is preferable to a lack of action caused by ignorance of those facts. Many companies dampen the effects of changes with rules like, "Don't reschedule open orders earlier within one week or later within two weeks."

Of course, not every change in inventory status calls for a reaction. Minor changes in requirements often are absorbed by inventory surpluses created by safety stock, scrap allowances, lot sizing, engineering changes, reduced requirements, supplier overshipments, early deliveries, and shop overruns. MRP uses up such excess inventories quickly through the net requirements replanning process. While inventory excesses are thus prevented from accumulating, they exist on some item or other at almost all times.

Planning safety stock, however, will not dampen nervousness, since MRP reacts to prevent the use of safety stock exactly the same way as it plans to prevent shortages. *The only effective way to avoid excessive nervousness from causing myriad corrective action signals is to eliminate the causes.* These include changes in demand data such as forecast errors, changes in customer order delivery dates or quantities, and revised schedules of warehouse orders. Many marketing and sales people believe that such changes are inevitable—"the customer is always right"—but others have found that better communication with customers reduces such changes by very significant amounts. In these companies, marketing/sales's primary goal is avoiding surprises from customers.

The other major source of nervousness is poor performance in supply activities. Reference has been made several times to the multitude of upsets in production that cause delays in completing orders. In the past these upsets were believed to be natural in manufacturing, and unavoidable. "If you can't stand the heat, get out of the kitchen," Harry Truman's famous comment about the U.S. president's office, is also the credo of rugged production people accustomed to living with crises.

Companies in every industry with world-class competitive status have shown clearly that such problems can be eliminated or at least reduced to negligible effects. Nervousness in MRP programs is a clear signal of poor performance in controlling demand and supply activities. It has long been clear that even powerful replanning programs cannot cope with such chaos.

PART 4

Looking Backward and Forward

It is the business of the future to be dangerous.

ALFRED NORTH WHITEHEAD

11
Lessons of the Past

Time...has taught us both a lesson.

PLUTARCH

Perspective

Before business computers became available, production and inventory control methods were ineffective. However, capital was available and relatively cheap, costs were not a serious problem, and customers were more tolerant of poor deliveries. Planning and control methods were crude, and constrained by primitive data processing tools (clerks and paper files) that were incapable of handling massive amounts of data. Theory and principles were lacking, and the only rigorous techniques known were machine loading (based on work standards, ca. 1900), economic order quantities (1915), and statistical safety stocks (1934). Ability to correlate multiple plans was nonexistent.

Business computers and software became available in the early 1960s and were applied by many companies to their planning activities. Forecasting, EOQ, and safety stock calculations were speeded up and made more frequently. Unfortunately the gains were minimal, at fairly high cost.

MRP programs, made practical by computers, were implemented in many firms beginning in the 1970s. Interest was greatly stimulated by an MRP Crusade conducted by APICS nationwide. The powers of the technique were stressed, but too little emphasis was given to the supporting activities of master-scheduling, structuring bills of material, getting accurate data, shop floor control, and capacity planning and control.

Little attention was given to integrating MPS and MRP into the total planning hierarchy that embraces

1. *Strategic planning*—long-range, broad-based, focusing on types of businesses, markets, and future directions
2. *Business planning*—long-range, focusing on product-family marketing and services
3. *Production planning*—mid-range, focusing on facilities, technologies, and all types of resources
4. *Master schedule planning*—short-range, defining specific end items, and driving detailed plans for resources

In many companies interfaces between these were weak at best, and often missing. Strategic plans, developed by top management, had little influence on production and master schedule planning. Marketing/sales business plans were largely ignored by production people. Their production plans, often used only for budgeting, were not linked to MPS. Precomputer manual systems were very difficult to link, and habits developed then were carried on long after powerful computers and software were available. Pre-MRP planning methods were not only fragmented and crude, they lacked completely an ability to replan, to respond to change.

More attention was given to integrated planning and control systems in the late 1970s, and to the task of educating people in the body of knowledge, language, principles, and techniques of manufacturing planning and control. APICS began to certify practitioners through a set of examinations covering these areas.

With the tools of planning and control understood and available, the 1980s focus was on solving the problems of execution. Long believed inevitable and unsolvable, the causes of upsets and interference with smooth, fast flow were attacked successfully by many companies. Unfortunately, too many others had already succumbed to competitors, most of whom were in foreign countries.

Computer-based MRP programs made possible six revolutionary advances:

1. Masses of data could be stored and manipulated cheaply, at blinding speeds.
2. Plans could be integrated over products and processes.
3. Complex product structures were easily loaded, stored, and retrieved for multiple uses.
4. User options were available for classification, lot sizing, safety stock, and other techniques.

5. Myriad data were available for many uses.
6. Frequent, rigorous replanning was possible easily, quickly, and inexpensively.

Early applications of MRP focused on Numbers 1, 3, 4, and 6; 2 and 5 were neglected, although these had great potential benefits and met many important needs.

MRP eliminated the greatest excuse for not executing the plan, that it was not valid. Overenthusiasm for MRP created the unfortunate impression that it was a system that could simultaneously improve customer service, reduce inventories, and cut manufacturing costs, along with performing other miracles. The truth, of course, is that *MRP is not a system, it is a technique* that can help people to do their jobs by

1. Recommending when orders should be released
2. Maintaining validity of order priorities
3. Providing data for planning capacity and other uses

The early promise of MRP was dimmed also by emphasis on using "standard MRPII systems." Self-proclaimed but underqualified system evaluators rated commercial software packages against the rater's specifications, which included many features inappropriate for some businesses. Repetitive manufacturing with no need for lot-order identity is an example. Reluctant to risk low ratings by not meeting all standard specifications, software suppliers included trivial features that added cost and complexity with little value to most users.

Implementation was viewed as "getting software running on the computer," not as using the programs to run the business, so users were poorly prepared, incomplete systems were installed, and proper foundations were not put in place. Underqualified consultants aggravated this by offering to help firms install MRPII systems and reach in very short time "Class A status," a set of superficial and systems-related criteria more than operations-oriented requirements. The quick fix appealed to many executives who lacked understanding of the requirements for sound planning.

The Real Problems

There are seven problems, one or more of which in the past have caused the failures of companies to gain the desired level of control of operations:

1. Poorly managed master production schedules, a topic covered in Chapter 10

2. Ineffective priority and capacity planning and control, discussed in several chapters
3. Inaccurate data, a principal topic in Chapter 8
4. Underqualified people lacking understanding of the principles and techniques discussed throughout this book
5. Upsets in operations interfering with execution of plans, covered in Chapter 10
6. Excessive lead times, much longer than actual work times, hamstringing sound planning, a major topic of this chapter
7. Poor organization, collections of all-stars with their own goals working at cross-purposes instead of as a team, discussed in Chapter 12

Rethinking Traditional Concepts

Long experience with MRP made clear the need to rethink certain traditional concepts and axioms in inventory planning and production control. It became obvious that most of these were no longer relevant or valid in an MRP environment.

Traditional ideas needing review related to

1. Use of ABC inventory classifications
2. Use of independent demand forecasts
3. Effects of demand and lead time errors
4. Effects of safety stock
5. Need for stockrooms
6. Controlling manufacturing lead times

Use of ABC Inventory Classes

The concept of ABC inventory classification, based on Pareto's principle of the vital few and trivial many in any set of variables, is presented in Chapter 2. In precomputer days, constraints on information storage and processing capacity made it impractical to give equal attention to all inventory items. Different planning and control techniques were selected to gain tight control of *A*-items, looser control of *B*-items, and only simple, crude control of *C*-items, avoiding shortages of low-value *C*-items by having plenty of them. "Having plenty" of *C*-items, of course, requires discipline in the use even of simple techniques.

Data processing limitations disappeared when computers became available and computer wizards deemed the ABC concept irrelevant; equal treatment was possible using MRP on all inventory items. Technically this is true, but several factors make users reluctant to include *C*-items under MRP programs:

1. There are so many of them—50 percent or more of all items.
2. Order quantities usually are large and orders are few.
3. Lead times often are short and many sources are available.
4. Precise physical control often is impractical.

Any MRP-planned item requires some user attention. Most MRP users prefer to carry more inventory of *C*-items, devoting planner time to more important duties; this is invariably true when new or radically revised MRP programs are being implemented. Low-cost *C*-items, especially purchased parts such as fasteners and small hardware, usually are ordered in large quantities and have short lead times; safety stocks are planned. Simple controls (two-bin, visual review) are adequate *if used well with discipline.*

Physical inventory control is an important factor in inventory management, but it is impractical to attempt to achieve precise physical control of many *C*-items. It simply is not worth the effort to try to keep exact counts of rivets, nuts, bolts, lock washers, and cotter pins; 95 percent accuracy usually is good enough. The ABC concept is valid also when applied to the activities of receiving, storage, issuing, and making cycle counts.

All items, including *C*-items, should be loaded into BoM so as to provide production people with complete information. When products are being assembled, all items should be issued in the right quantities at the right time. Even simple control techniques need data on annual usage of *C*-items; bill processor programs can provide these easily and quickly via summarized explosions.

It is true that MRP control, when feasible, can help to improve coverage and to reduce inventories of *C*-items; scarce and expensive capital is wasted on any unneeded inventory. Priority planning cannot be as good when all manufactured items are not covered by MRP; manufactured *C*-items often compete with *A*- and *B*-items for scarce capacity and other resources; tooling and test equipment, for example. Purchased *C*-items usually do not compete with other purchased materials and pose little problem if not in MRP.

Capacity requirements data will be incomplete if manufactured *C*-items are excluded from MRP. And not only their absence will cause errors. Simplified control may make the scheduling of their operations incorrect, thus affecting the validity of the entire load projection and capacity analyses.

A useful approach to getting most of the benefits of MRP coverage with the least effort is to make an ABC analysis of *C*-items, defining the classes by whatever criterion is important. If it is better priority, for example, a *C-A*-item would be one processed in critical work centers; if capital were tight, the *C-A*-class would include the *C*-items in the top 20 percent of usage value.

Forecasting Independent Demand

MRP calculates dependent demands (from parent needs), determining how many and when components of parent items are needed. Independent demand (from external sources) must be forecast. This classic problem of inventory management is greatly simplified and made easier in modern planning and control. Pre-MRP methods attempted to forecast demand for all items, whether independent or dependent. No forecast is required, of course, when materials can be acquired and converted into products within the delivery time allowed by customers. Unfortunately there are few such products now, and likely to be fewer in the future.

When advance commitments have to be made to procure or manufacture items, forecasts are necessary. Forecast accuracy is an oxymoron; attempts to predict future events will fail. Extensive (and expensive) attempts to develop sophisticated forecasting "models" of markets for manufactured products have been little better than human intuition, often not as good. *Forecasts always will be wrong.* This is the first principle of forecasting. The important question is "*How* wrong?" The answer is itself a forecast, but the assumption that future errors will continue to be like past errors is often valid and easily tested.

Making accurate forecasts is far more difficult than using poor forecasts well; learning to use poor forecasts better has real potential benefits. The less flexibility there is in subsequently modifying the original plan, the more important is the accuracy of forecasts. Conversely, when replanning is easy, valid, and continuous, forecast accuracy is of much less importance *if production is flexible and can react quickly to changes in plans.* Flexibility in replanning and execution are the keys.

Effects of Demand and Lead Time Errors

TPOP and MRP replan to react to both "bad" forecasts and "inaccurate" lead times. Pre-MRP techniques were unable to replan, and depended on the accuracy of demand forecasts and lead time data, unable to adjust for errors in variables inherently volatile. MRP and TPOP use forecast demands and planned lead times as points of departure, these data serving as raw materials for the construction of a "live" plan that is modified to reflect reality. These techniques adjust to what is happening; they are not restricted to what was planned to happen.

These are not unmixed blessings, however. Demand forecasts and planned lead times must be viewed as averages; actual data will always vary above and below them. Planning techniques that "adapt" to actual data rapidly and frequently can become very nervous, sending signals of needed changes that can overwhelm suppliers and work centers. Many of these replanning signals will cancel each other out over time, even if average forecasts and lead times are correct, as actual data oscillates above and below them. Excessive planning nervousness is a serious problem; this topic is discussed in Chapter 10.

Effects of Safety Stock

Safety stocks of raw materials and components have always been part of stock replenishment methods. Order point and cruder techniques used in the past depended on such cushions of extra inventory to compensate for their inherent deficiencies as well as "the slings and arrows of outrageous fortune" that frustrated manufacturing operations. This thinking carried over into MRP, which can easily incorporate safety stock in planning (see chapters 4 and 6).

It has little, if any, legitimate role in MRP, however, which provides forward visibility of planned requirements and orders for raw materials and components and can replan easily. *When safety stock is added to MRP, the resulting overstated requirements and false timing of order release and due dates destroy its credibility.* Factory personnel quickly discover lack of integrity of priority information and soon learn to disregard order due dates, believing that safety stock is available and that orders are scheduled for completion before they are actually needed. Missed delivery dates without customers' quick and strong reactions give suppliers clear evidence that their purchase order due dates are padded.

The primary purpose of safety stock is to provide cushions against fluctuations in demand (forecast errors) and supply (upsets in production). Demands for raw materials and components are calculated by MRP and, therefore, are not subject to forecast errors. Experience with safety stock in MRP has shown clearly that the cushions are rarely found on items affected by upsets, nor are they large enough to be useful. Better understanding of the importance of smooth, fast flow of materials and flexibility of operations led to the realization that the causes of upsets had to be eliminated. Efforts to do this proved successful and yielded enormous benefits far beyond the costs incurred.

Safety stock is properly applied only to inventory items subject to independent demand. These include stocked finished products, service parts, and MPS end items. This topic is covered in Chapter 6. Safety stocks at the MPS level protect MRP against both forecast errors and nervousness, but they need not and should not be duplicated at raw material and component levels. Safety stock at the MPS level has the added advantage of planning extra sets of components *in matched sets* required to build end-item safety stocks. This will never happen when component safety stocks are planned independently of each other.

Another function of safety stock is protection against uncertainty of supply. It can be useful sometimes on an item whose supply is erratic and beyond control of people in the plant, characteristic of some purchased items. Safety stock will be wasted when production of manufactured items is erratic and unpredictable; planning cannot be sound in chaotic environments.

The venerable concept of safety stock has been rethought. Modern inventory management utilizing MRP and TPOP applies their abilities to realign open-order priorities when requirements change, and understands that lead times are flexible. Planned lead times are not fixed realities, they are averages. Actual lead times can be compressed or extended by raising or lowering order priorities. Safety stock is not needed in well-run plants to cover upsets during planned replenishment lead times, with dynamic replanning coupled to flexible, reactive execution.

MRP and time-phased order points (TPOP) can replan quickly, accurately, and automatically. TPOP, particularly, replans with equal ease whether needed for revisions of forecasts or differences between forecasts and actual demands (errors). The self-adjusting capability of this technique makes reasonable forecast errors unimportant in flexible operations. TPOP needs forecasts of demand to function, but it does not depend on their accuracy for its effectiveness.

Time-phased order point provides an excellent example of an inventory planning and control technique that "works" almost independently of the quality of the forecast. Figure 11-1 shows an item's inventory record under TPOP; section A has gross requirements based on a forecast of 30 units every period. The quantity currently on hand is 140, and planned safety stock is 15. On-hand is projected to be below this level in period 7, so the next planned replenishment order is scheduled for release in period 4, allowing 3 periods of lead time.

If actual demand in the first period is zero rather than the 30 forecast, section B of Fig. 11-1 shows the item's status from period 2 through 7.

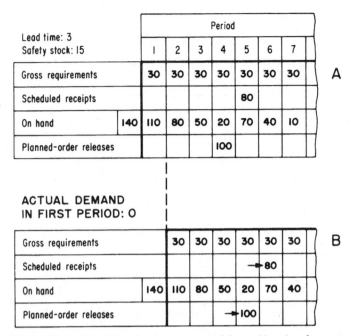

Lead time: 3
Safety stock: 15

		Period						
		1	2	3	4	5	6	7
Gross requirements		30	30	30	30	30	30	30
Scheduled receipts						80		
On hand	140	110	80	50	20	70	40	10
Planned-order releases					100			

A

ACTUAL DEMAND
IN FIRST PERIOD: O

Gross requirements		30	30	30	30	30	30
Scheduled receipts						→80	
On hand	140	110	80	50	20	70	40
Planned-order releases					→100		

B

Figure 11-1. Time-phased order point: actual demand less than forecast.

Projected on-hand inventory data are all 30 units higher than in section A, the scheduled receipt (open order) is not needed until period 6, and the planned order now can be released in period 5, one period later than previously planned. This will affect net requirements and coverage. If there is again no demand in period 2, these orders can again be moved back; if demand equals forecast, the orders will stay as presently scheduled; but if total demand in periods 2, 3, and 4 exceeds forecast by 6 or more units, TPOP will recommend advancing them.

If actual demand turns out to be 90 in period 1 instead of the forecast 30, the quantity on hand at the end of period 1 will be 50, as shown in Fig. 11-2, not the previously projected 110, changing both the open-order and planned-order schedules. The date of need of the open order advances to period 3, and the planned order should be released now.

Actual demand for independent-demand items can be expected to keep changing; forecasts, at best, may equal average actual demand but rarely will be accurate for individual periods. No matter how large forecast errors may be, TPOP adjusts plans automatically when notified of actual demands.

This self-adjustment characteristic of TPOP is particularly significant for maintaining true priorities of open orders. In these examples, if actual demand drops to zero (the item becomes obsolete), the open-order

Lead time: 3
Safety stock: 15

		Period							
		1	2	3	4	5	6	7	
Gross requirements		30	30	30	30	30	30	30	
Scheduled receipts						80			
On hand	140	110	80	50	20	70	40	10	
Planned-order releases					100				

A

ACTUAL DEMAND
IN FIRST PERIOD: 90

Gross requirements		30	30	30	30	30	30	
Scheduled receipts			80 ←	←				
On hand	50	20	70	40	10	-20	-50	
Planned-order releases		100 ←	←			100 ←	←	

B

Figure 11-2. Time-phased order point: actual demand exceeds forecast.

need date drops back continuously (indicating that work should be stopped) and no planned order will be released. Early-warning signals of imminent obsolescence (such as no demand in 4 periods) would limit the costs accrued.

Need for Stockrooms

MRP logic described in Chapter 4 makes an assumption that each item under its control passes into and out of stock, and that transactions reporting receipts and disbursements will be generated. It is not practical to route every inventory item through a stockroom in many manufacturing operations, however. In such cases transaction reporting, mandatory under MRP, may be based on events other than physical arrival in and departure from stock. Transactions may initiate from

1. Reports from stockrooms
2. Reports from receiving docks
3. Shop floor events
4. Other transactions

Reporting from the stockroom is the normal practice, but often receipts of purchased items come from *receiving docks*. When stockrooms are bypassed, transactions can signal both a receipt and a disbursement. Receipts and disbursements can be tied to *shop floor events*; for example, completion of the last operation on a work order may be considered a receipt of the item covered or even a simultaneous receipt and disbursement to another production stage. Completion of a parent order may be treated as disbursement of its component items, and assembly-line production reports may be translated into many component disbursements. *Some transactions may trigger others*; for example, releasing a parent planned order may signal disbursement of its components.

There are several potential dangers in divorcing physical movements of material from data movements. Finding record errors and determining their causes is made more difficult, at best, and often is impossible. Control of material movements to their proper destination is not as tight. Delivery of matched sets of components to assembly areas is not practicable. In addition, stockrooms often perform useful minor operations preparing materials for production.

There are strong reasons for eliminating stockrooms. Modern just-in-time factories with flow operations like process plants have "point-of-use storage" on the plant floor for small quantities of raw materials and components. Based on the Japanese Kanban technique, batches of material in standard containers are planned to supply each operation. Use of some material from one container triggers replacement actions of supplying operations without paperwork or data transactions. Literature in the Bibliography contains details of this approach.

Controlling Manufacturing Lead Times

There are two types of lead time used in manufacturing planning, *operation* lead time and *order* lead time:

1. *Operation lead time*—the average time required for all activities to complete work on one operation on an order.
2. *Order lead time*—the average time required for all activities to complete work on an order for one lot of material. Order lead time is the sum of all operation lead times on an order.

The relationship between these two, and their use in controlling priorities, is described in Chapter 10. Order lead times are set by planners and loaded into MRP databases. Operation lead times are calculated from order lead times by scheduling techniques that use rules for each lead time element.

There are three types of order lead times:

1. *Planned*—the time between start and completion of work orders, used by MRP for planning and replanning
2. *Actual*—the time experienced by orders moving from start to finish during execution of the plan
3. *Available*—the time remaining before orders are needed

Planned lead times are established by planners and become part of each item's master record. They are estimates of the average time an order for each item will take to move through production operations, and they must include allowances for each of the four elements. Operation lead times have four basic elements:

1. Setup time—required to change over machines or work stations from making one item to making another
2. Run time—required to process the order lot
3. Move time—required to move between operations
4. Queue time—waiting between operations

Surprisingly often, queue time accounts for 90 percent or more of order lead time. Real work is involved only in the first three elements, setup, running, and move times. Compressing queue time by reducing work-in-process is the most effective way to reduce overall lead times quickly; lead times only 10 percent or less of original values can result when queues are compressed.

Setup, run, and move time, the working elements, account for only a small fraction of total lead time, but it is a mistake to believe that reducing them can bring only small benefits. Shorter setups permit the running of smaller lots without added costs, and also increase work center capacity. Small lots reduce investment in work-in-process, cut lead times, and aid greatly in smoothing the flow of materials through operations. In addition, small batches can be moved easily and quickly between operations, often by hand by machine operators, eliminating the need for material handling equipment and move delays.

Making planned lead time estimates is not a simple matter of averaging the last three or four actual lead times required to complete orders. Over a reasonable period of time, orders will require a wide range of time to move through most work centers. Figure 11-3 shows that some orders passed through work center 103 in as little as 3 days, while some took more than 10.

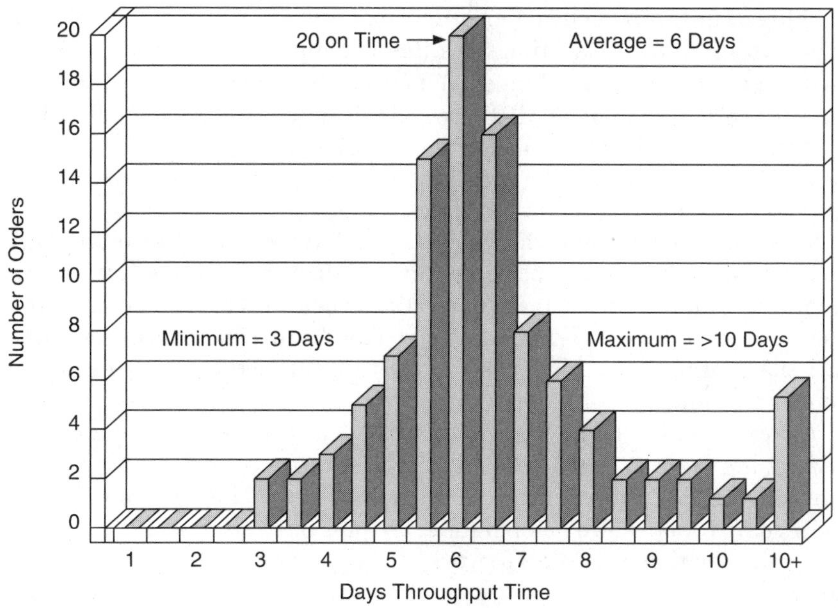

Figure 11-3. Lead time variations in work center 103.

Why didn't any orders take less than 3? Probably because of long setup and run times and/or long move time. Why did some take more than 10? Undoubtedly because they had low priority and were allowed to sit in the queue. What actual lead time should be used in MRP for this work center? The greatest number of orders took 6 days; this is good enough to use at the start of MRP.

A little-known (and difficult to accept) fact about lead times is that *they will be what you say they are!* Setting planned lead times lower than current average actual values (don't get reckless and cut them in half at once) will shut MRP off from recommending the release of new orders for the period that lead times were cut. Work centers will work on the orders already in work-in-process and the queues will shrink, making actual lead times shorter. MRP responds quickly to changes in planned lead times.

Planned lead times must never be increased, however, no matter how much actual lead times exceed them, making deliveries late. This triggers the Lead Time Syndrome, a vicious cycle. If planned lead times are increased for production operations, for example, MRP will immediately signal the need to release more orders, these will add to queues, actual lead times will get longer and more erratic, and more orders will be late.

Purchased materials also will experience a similar vicious cycle if planned lead times are increased. For example, if a supplier sees that its backlog of unshipped customer orders exceeds its ability to deliver all of them within quoted lead times, it often is tempted to inform customers that lead times now are longer. When these increased lead times are inserted into a customer's MRP program, it signals the additional orders to go to the supplier, who thinks that "business is picking up and good times lie ahead." In fact, the actual load on its facilities is larger now, and quoted lead times are again too short.

Repeated trips around the vicious cycle have occurred in many U.S. industries; machine tools, bearings, semiconductors, and steel are prominent. Customers soon learn, of course, that other sources are available, and they change suppliers. Those who played the Lead Time Syndrome game lost share of market; this has been a significant factor in the decline of American industry.

The only solution to the problem of actual lead times exceeding planned is additional capacity to work off the excess load. Overtime, added (temporary) workers, outside subcontracts, running smaller batches (if not offset by more setups), and alternate operations are possible ways to increase output.

During the time this takes to work off excess queue, it is often better to cut planned lead times. This shortens actual lead times (as described earlier in this section), improves the validity of priorities, and limits chances of working on the wrong orders. It is extremely difficult, however, to get production people to understand and accept the fact that this actually will happen.

Actual lead times vary widely, as Fig. 11-3 shows so clearly, even for successive orders for the same item, since they depend on order priorities. The higher an order's priority, the shorter its actual lead time will be. Priority determines the sequence in which orders are ranked. No two orders can have the same place in the sequence; one must take precedence over others.

Available lead times change with the passage of time and changes in due dates (established when orders are released) when MRP replans, setting new need dates (see Chapter 5). Earlier need dates means less time to finish orders and give them higher priority than others with later need dates.

Compressing actual order lead times rarely requires splitting orders; almost always it can be done by bypassing other orders in queues of work-in-process, occasionally also expediting moving orders between operations. The lead times of only a small number of orders can be compressed significantly at any one time; obviously, everything cannot be expedited at once. The ability of work centers to collapse lead times of specific orders depends on two factors: first, how many orders require collapsed lead times; second, actual capacity of the work centers.

Computer programs called *operation sequencing* (the original name was *finite loading*) develop detailed schedules recognizing capacity constraints and limited flexibility. These are quite different from those based on MRP need dates, which

1. Ignore capacity limitations
2. Assume average lead times for parents and higher-level items
3. Plan full order quantities
4. Don't consider lap-phasing orders moving through successive operations
5. Don't plan to run orders on two or more machines simultaneously
6. Ignore adding capacity by overtime, alternate operations, or subcontracting work outside

These programs are not alternatives to MRP; they work closely with it, refining its data and applying it to enhance execution of its plans. They have been applied successfully in many companies. They work best where a few bottleneck machines are critical to schedule adherence.

Load versus Capacity

Confusion often exists as to the meaning and use of the terms *load* and *capacity*. As defined in Chapter 8, load is the depth of liquid in a tank such as the one in Fig. 11-4; capacity is the rate at which it is running in (input capacity) and out (output capacity). Queues of work-in-process are visible loads. Open-order and planned-order schedules in MRP are easily converted by capacity planning programs to load reports, using the process described in Chapter 9. Qualified people know how to read and use them correctly.

Underqualified people confusing load data with capacity requirements make frequent mistakes in identifying true problems. Late orders are seen as priority problems, when the cause more often is inadequate capacity. High levels of work-in-process and late orders are believed to indicate capacity problems but are often caused by excessively long planned lead times. Using MRP programs and their outputs effectively requires a sound understanding of both load and capacity.

In many companies, overloads coupled with heavy behind-schedule loads appear to be a permanent condition. At the same time, shipments of products are reasonably on schedule. Such load data are viewed with healthy skepticism, and people are loath to act on the information provided. This situation constitutes virtual proof that load reports are

1. *Incomplete*—fail to include load that will be generated by planned orders
2. *Invalid*—priorities are not up-to-date

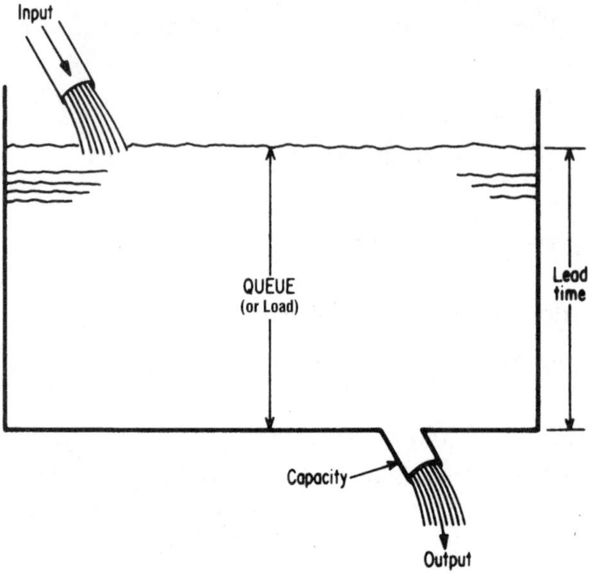

Figure 11-4. Analogy of queue and water in a tank.

Including only open (released) orders in load reports causes indicated loads to decline over short horizons equal to the span of the average item lead time. This provides little "visibility" into the future beyond the current period and defeats the purpose of projecting the work load. Capacity corrective actions such as hiring or subcontracting require significant lead times; the very information most desirable, a valid load picture reasonably far in the future, is missing.

If plans are to be valid, average actual lead times of successive orders for any item should equal its planned lead time (also an average). The former is a function of actual capacities and average levels of work-in-process in all work centers involved in processing the item's orders.

Large "bulges" in behind-schedule and current-period load, referred to as *front-ended,* often are indicators that priorities are not valid. Portions of the load classified as "behind schedule" may not really be late. Requirements may have changed, but order due dates and operation dates have not been revised to reflect this. Some of the overload in the current period also may be false.

Before loads can be evaluated properly, true priority of orders must be known. Figure 11-5 is a modified tank, like the one in Fig. 11-4, in which order priorities have been used to stratify the queue. A drain has been added to draw off dead orders (no requirements) that may appear. The pump moves high-priority orders promptly. Failure of the work center to

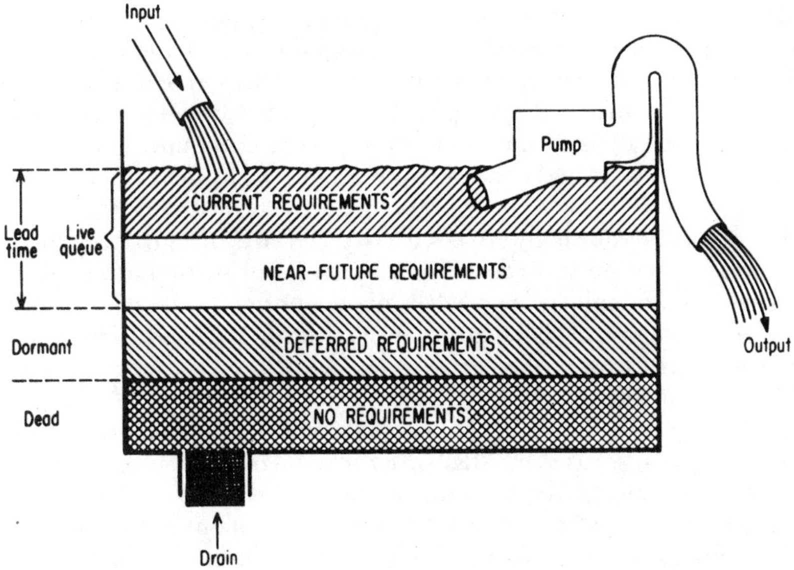

Figure 11-5. Queue with priority strata.

process enough orders in priority sequence to *prevent steady increases in queue* is clear evidence of inadequate capacity. The mere presence of a queue larger than the work center capacity is not.

Some behind-schedule (late) orders also are not proof that capacity is too small. Operation schedules are arbitrary prorations of order schedules over several work centers, and contain allowances for waiting in queues. Late orders in one work center, processed promptly, may move to another, be worked on soon, and actually get ahead of schedule. Load and priority data for one center do not provide enough information to identify real problems and corrective actions; downstream conditions are relevant also.

When capacity is inadequate, planners must determine answers to difficult day-to-day operating questions like

1. Should we work overtime?
2. Should we transfer work from one department to another?
3. Should we transfer people from one department to another?
4. Should we subcontract some work?
5. Should we start another work shift?
6. Should we hire more people to work current shifts?

MRP does not itself plan capacity requirements, but it provides vital input to capacity requirements planning programs (see Chapter 8). It also can be useful in capacity control activities (see Chapter 10), providing data on individual orders that might be transferred between work centers or subcontracted. It is of little help on capacity, economic, and policy information needed to make the tough choices posed by the preceding six questions.

Load projections normally are not perfectly level; actual loads fluctuate from period to period, as orders requiring larger or smaller amounts of work arrive in random patterns. MRP does not ensure level rates of work release, even when care is taken to develop level MPS. Lot sizing and varying mixes of orders flowing through work centers cause fluctuations in period loads.

Capacity decisions, however, can be made with reasonable confidence in the average loads, based on MRP order schedules. Short-term capacity adjustments are necessary to compensate for load fluctuations from period to period. The last section of this chapter contains a discussion of the use of input/output control to minimize these effects.

Queue Analysis and Control

Queue analysis and control are based on valid work order priorities. Queues, as shown in Fig. 11-5, can be stratified into "live" (current and near-future), "dormant" (deferred), and "dead" (no requirements) segments using dates in MRP order schedules. Orders in queue have two kinds of priority: relative and absolute. Figure 11-6 illustrates these for six orders in the current load. The traditional (left-hand) view is that every one of these is eligible, in priority sequence, to be worked on in the current period, but this is not necessarily so.

Valid queue analyses cannot be based on relative priority alone; absolute priority also must be considered. Relative priority is simply a ranking of a group of orders; absolute priority considers each order's need date. The right-hand view of the blocks in Fig. 11-6 shows that the active queue contains only one-half of the orders.

Conventional queue control before MRP programs were used involved measuring queues at a work center at intervals over a period of time. Results might be as shown in Fig. 11-7; a maximum might be 100 and a minimum 60 standard hours. It was assumed that 60 hours was excess above the minimum queue required to prevent the center's running out of work, and action was taken to remove this excess through overtime, subcontracting, or transfer. The "controlled" queue was then assumed to fluctuate between 0 and 40 standard hours.

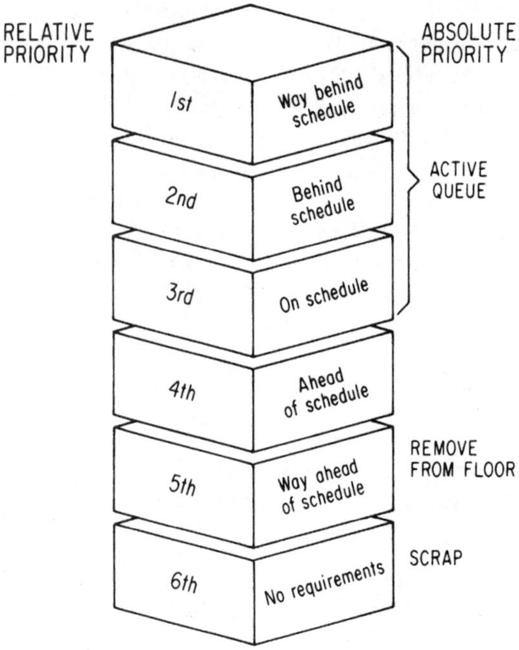

Figure 11-6. Relative and absolute priority.

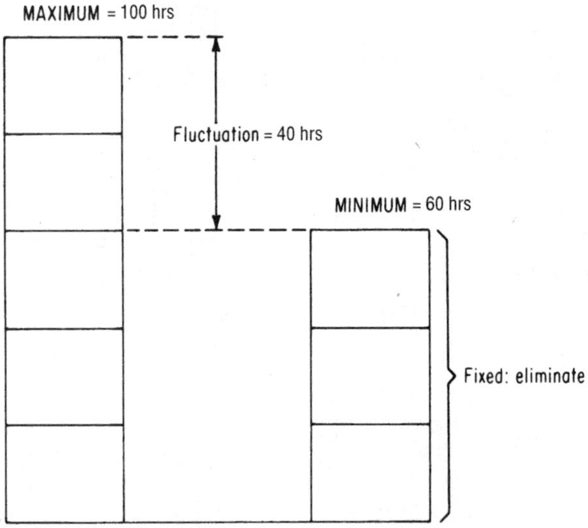

Figure 11-7. Queue control: conventional view.

Unfortunately, standard hours of work do not describe a queue adequately. Units of work are not necessarily homogeneous and interchangeable. Working off 60 hours would certainly include many orders with high relative priority, and the queue would be reduced from the "top" rather than from the "bottom," as shown in the figure. This approach, at best, is impractical; spending extra money and effort to reduce this queue may simply move the excess downstream to wait in another center. Spending extra money to leave only low-priority orders to be worked on makes little sense. This example shows how simplistic solutions to complex problems usually give unsatisfactory results.

Input/Output Control of Work-in-Process and Lead Time

Work-in-process and lead time codetermine each other. Work-in-process inventory, lead times, and output rates have been related traditionally by the following formula:

$$L = \frac{W}{R}$$

where L = average lead time (days, weeks, or months)
W = work-in-process inventory (units, hours, or dollars)
R = rate of output (per period of L, in units of W)

For example, if $W = 1200$ units and $R = 200$ units per week,

$$L = \frac{1200}{200} = 6 \text{ weeks}$$

However, average lead time calculated this way may not be meaningful. Average actual lead time is a function of the live portion of work-in-process; the dormant and dead portions have no effect. When work-in-process is stratified by priority, the results of applying the formula are different:

Active queue:	800 units
Deferred requirements (dormant):	300 units
No requirements (dead):	100 units
Total	1,200 units

For this case,

$$L = \frac{800}{200} = 4 \text{ weeks}$$

The difficulty in using this equation is that it assumes that work-in-process and lead time are relatively fixed. In reality, of course, WIP is variable; its level depends on the relationship between period work input to and output from WIP. When input exceeds output, WIP and lead time go up, and vice versa. In order to reduce WIP and lead times, output must exceed input. Control of WIP and lead time requires a proper balance of work input and output—total work entering and leaving work centers.

Work orders recommended by MRP are released by planners into starting (gateway) work centers. As discussed in Chapter 10, MRP does not recommend level loads for release (Fig. 10-8 shows this clearly), but planners can regulate input of orders to maintain a level load by releasing some orders early and holding others for later release. Order release must include a check of components and other resources to ensure that no shortages exist.

Many underqualified people object to overriding MRP recommendations in the belief that it states the correct priorities in the correct sequence and at the right time, and that priority control is more important than work-in-process control. They overlook four factors:

1. Planned lead times in MRP are averages; there is nothing magic about recommended order start dates.
2. Control of average lead times requires WIP control.
3. Releasing more orders than a center can work on gives real control of priorities to shop people.
4. Good performance and customer service require that plants finish on time, not start on time.

Without tight planner input control, a significant change in requirements resulting from revised MPS will allow MRP to pump out a large number of new orders on top of existing work-in-process. The revised MPS may meet marketing needs and also be realistic in terms of capacity requirements, but even temporarily increased WIP levels will cause longer, more erratic lead times and poorer priority control.

Input regularly exceeding output is a clear indication that MPS are overstated and capacity is inadequate. A strong case is made in Chapter 8 that this situation is very dangerous; *MPS should be reduced if capacity cannot be increased promptly*. Since the latter action is difficult and often takes considerable time, managers often view increased MPS as "temporary" and assume (hope!) that the problem will soon go away and that "a little overtime" will handle overloads. The harmful effects of even such wishful thinking can be minimized by tight input control.

Planners must walk a narrow line between following MRP recommendations blindly and overriding it willfully. Excessive caution by planners may cause serious problems. An example of problems caused by releasing a work order for 40 pieces instead of the 25 pieces planned is given in Chapter 9, page 220. As shown, this override of the MRP plan can generate serious problems for component work orders already issued, for work centers with heavy loads, or for other parent items using the same components. Experienced planners realize that attempts to get "cushions" in this manner can "backfire" badly. There may be good reasons for increasing the order from 25 to 40, of course, and experienced planners will attempt to minimize the problems they know will be caused.

If the item is purchased, the same problem may occur at the supplier's plant. Modern partnership relationships with suppliers (discussed in Chapter 12) include full communication of customer requirements and lean inventories in both parties' plants. Arbitrary changes such as the one in this example will quickly destroy good working relations.

Managing Service Parts Inventory

Historically, managing service parts inventory evolved toward organizational separation and independence from the manufacturing supplying function in general and production and inventory planning and control in particular. This was caused primarily by marketing interest in service parts and by the need for market orientation of the people in charge of this inventory. The first step in this evolutionary process was to segregate service-part stocks from production materials physically; the intent was to prevent production personnel from "borrowing" service parts to cover shortages caused mostly by deficiencies in plant inventory control. Replenishment orders were issued by marketing or repair parts people, usually passing through plant inventory control.

Next a separate service-part department was set up, independent of plant inventory control, and it developed its own methods. This generated as many problems as it solved and even greater separation was made; service-part organizations began to operate their own warehousing and distribution facilities geographically remote from the manufacturing plant.

Separate ordering and stocking of service parts caused many serious problems. Orders had to be placed on plants using standard lead times but covering requirements over much longer times; plants needed these data for capacity planning. Orders covering unexpected demands were entered with less than normal lead times, but unneeded orders rarely were rescheduled later. Plant people gave higher priority to plant needs, and service part shortages were common, causing growing backlogs of delayed customer orders. Periodically, top management directed concentrated efforts to reduce these, and the pendulum swung the other way to increased plant shortages. The underlying problem was failure to integrate planning and execution for both types of inventories.

Implementation of MRP reversed this trend toward separation and independence. First came a recognition of the benefits to be gained from uniting inventory planning for service and production parts, using TPOP for the former and MRP for the latter. This gave plants a clearer view of future service-part requirements, improved planning and control of both inventories, and lowered total inventory investment.

Additional benefits were gained by consolidating the bulk of these inventories. Parts common to both could be allocated to production and service-part needs when demands firmed; this made the best use of available inventory, yielding improved customer deliveries of service parts and fewer production shortages. Remote warehouses for service parts then carry only minimum amounts shipped on short notice from central plant stocks.

12
The Future of MRP

In the past we had a light which flickered, in the present we have a light which flames, and in the future there will be a light which shines over all the land and sea.

SIR WINSTON S. CHURCHILL

A Brief History

The introduction of computers into production and inventory control in the mid-1960s provided orders-of-magnitude increases in the power of available techniques. These resulted from the enormous power of computer hardware to enter, store, and retrieve data and to manipulate the masses of data found in even the smallest manufacturing businesses. Coupled with basic software such as bill-of-material processing, inventory status, master scheduling, materials requirements planning, and capacity requirements planning, it was then possible to solve old problems, even those that in the past had been the most baffling and stubborn.

Further refinements and new developments in software came with subsequent generations of computer hardware that multiplied power while dividing costs. Computer wizards believed that more sophisticated planning programs could solve all manufacturing problems. Billions of dollars and millions of hours were spent implementing and installing complex programs, most of which never achieved more than a fraction of potential benefits.

There are four principal reasons (and a host of lesser ones) why MRP failed in the past to deliver full potential benefits:

1. *Mismanaged master production schedules.* The serious, sometimes fatal, problems of overstated MPS overloading plants and suppliers are covered in Chapter 8. The harmful nervousness resulting from "tinkering" with MPS is discussed in Chapter 10. Management's failure to understand the role of MPS, to resolve conflicts among organization groups, and to insist on realism hamstrings MRP in making valid plans.

2. *Inaccurate data.* Significant errors in bills of material, inventory balances, and open orders preclude MRP from producing valid outputs. Input data integrity is discussed in Chapter 8.

3. *Improperly structured bills of material.* Planning BoM, different from engineering, production, and cost bills, greatly improve the quality of plans, particularly for complex products with customer options. This is covered in Chapter 7.

4. *Lack of user control.* Inexperienced planners, who don't know how to read output reports or foresee the consequences of overriding MRP, naively accept computer recommendations. Lack of credibility of MRP outputs causes many capable planners to refuse to use them, depending on side records and subsystems. Failure to take important actions, such as rescheduling open orders, negates the benefits MRP makes possible; an excessive work load is often the reason for such failure. These topics are discussed in Chapter 10.

Early Research

Early research into material requirements planning and related topics focused on four areas:

1. *Techniques for calculating economic order quantities.* This work produced the many variations on EOQ presented in Chapter 6. Occasional writings still appear on "EOQ Revisited" themes, but no more work on this topic is needed. *The ideal lot quantity is what customers want today,* and setup times can be shortened to make such quantities feasible in most companies.

2. *Use of safety stock.* Statistical forecasting and analyses of forecast errors fascinated early MRP users, and efforts were made to develop software to link these to safety stock cushions in master production schedules and MRP. This was not only futile but counterproductive. As discussed in Chapter 11, safety stock in MRP is now recognized as

much more of a hindrance than a help. It violates the basic strategy of not committing flexible resources to specific items until the last possible moment.

3. *Structuring bills of material.* Chapter 7 presents the development and use of modular BoM for complex products with optional features. Concurrent engineering, teamwork of design engineers and others involved in planning and execution, is making modularity an objective. This topic still needs some attention but it is not a high-priority item.

4. *Scheduling work orders.* Much work has been done in academia and consulting firms attempting to apply mathematical formulae to shop scheduling. Called *operations research,* this invented techniques (algorithms) based mainly on linear programming and queuing theory. Mathematically elegant, many were solutions to trivial or oversimplified problems that could be solved better by more practical actions.

People working in or closely related to industry frequently view research efforts by educators and consultants as "those having a few familiar tools running around looking for something to fix" rather than seeking solutions to real problems. Researchers should address what needs to be researched, not what is intellectually challenging, and educators should teach what people need to know, not what is eminently teachable.

Closing these gaps can come only from better cooperation and dialogue between academia and industry. The certification program of the American Production and Inventory Control Society has aided in improving communications between them. The inhabitants of the "real world" must in the future take the initiative in articulating and communicating their problems to researchers and educators, and they should actively support valid research. Researchers, on their part, should validate their research programs before undertaking them by asking the real world they hope to assist, "Is this one of your more pressing problems, and should we be working on it?"

In the early 1980s, experienced and qualified people in manufacturing planning and control recognized that good computer-based systems were necessary but not sufficient for success. Planning, however sophisticated, is futile in an environment out of control. Instead of planning cushions against upsets, and replanning early and often to react to them, problems causing delays in operations had to be solved.

That this was possible and was being done was proved by some real successes in the United States and Japan. The former received little publicity, but politicians, media people, and consultants trumpeted the Japanese story, relating the supposed reasons for their recent domination of many industries—radios, televisions, motorcycles, and many more—and deep inroads into others—automobiles, machine tools, and heavy machinery—that formerly were American bastions.

By the early 1990s, it was clear that there was no mystery or magic involved in becoming a world-class competitor in global markets. The keys are total quality management, high data integrity, sound planning and control, fast and flexible production, and teamwork among organization groups including customers and suppliers. MRP has an important role to play as the central technique in the core system. While all of this is known, however, much work needs to be done to understand fully how the keys to world-class performance are implemented in industry and how people can be convinced that the principles are right and the techniques useful in their own operations.

Effects of Just-in-Time Operations

MRP originally was applied to operations using discrete orders to meet requirements, and there will always be many firms who must operate this way. The dramatic improvements in operations from smoothing out and speeding up flows of material, together with elimination of all kinds of waste, have resulted in the adoption of Just-in-Time (JIT), Total Quality Management (TQM), and Concurrent Engineering (CE), all designed to move batch operations toward process-industry flow.

Some believe that these replace MRP, but this is wrong; MRP basic logic is still applicable. Significant changes are needed, however, in the way MRP programs handle data:

1. Speedier operations on smaller lots need shorter periods in execution; days, even hours, replace the weeks commonly used.
2. Paperwork (order packets) are not needed to authorize work; the presence of the material is enough.
3. Rate-based planning replaces discrete orders, simplifying data needed. Lot sizing is stated in units per period, not per order.
4. Tracking batches is unnecessary; exception reporting of delays is sufficient.
5. Multiple-location, on-hand-balance fields are required for components stored in production cells, not stockrooms.
6. "Backflushing," reducing component inventory balances by computer calculations from parent reported production, replaces individual issues and allocations.

7. Cost accounting must change from order-related to time- or process-related.
8. Traditional batch MRP programs may still be needed for low-volume products built intermittently.
9. Planning level rates in weekly periods over longer terms is simpler, with little need for net change MRP programs.

Capacity requirements planning is both simpler and more critical with JIT operations. Rough-cut CRP using bills of labor for families of items running through cells or machine groups is adequate based on rates of throughput. Input/output control also is simpler; much of this activity is done in planning the cells and balancing machine rates. Flow-line operations, Kanban, total quality management, concurrent engineering, and other recent developments are enhancements to MRP, not replacements.

The Future of Material Requirements Planning

A manufacturing business is a living organism. The world economy is changing continuously, and success in global competition requires constant, unrelenting efforts to improve all activities, speed them up, and make them more flexible. Talk of the United States moving from manufacturing to a postindustrial, high-tech, or service economy is based on the fallacious assumption that these can generate the real wealth required to provide an adequate standard of living for a growing population. They cannot. Competitive, broad-based manufacturing is essential.

The process of improving manufacturing is well known. A fundamental part of this is sound planning and control, of which MRP is an essential element. How it works and how it helps to make manufacturing work better are the subjects of this book. The compelling need to improve execution has been stressed.

When the manufacturing environment is cleaned up and streamlined, planning is greatly simplified. Products designed for good manufacturability have fewer components, simpler bills of material, lend themselves better to planning, and have fewer engineering changes. Materials that flow smoothly need no inventory buffers or stockrooms and do not have to be tracked and reprioritized. Fast, flexible operations avoid using valuable resources to make items not needed soon, and shortages of adequate resources to make what customers want now.

Many new concepts, techniques, and computer-based systems to apply them have evolved over the past three decades. Most of the pre-MRP techniques and approaches have diminished in use but still find some good applications. Professional manufacturing planning and control people know how to blend new and old. A long time ago, a real professional named W. Evert Welch said, "There are no bad techniques in our field, just bad applications of good ones." This is as true now as when he spoke.

MRP will not become obsolete. The logic of MRP, set forth in detail in Chapter 4, is the fundamental logic of manufacturing as set forth in Chapter 2. It has been and will continue to be an integral part of the core planning and control system, regardless of what products are made, how they are produced, where the plants are located, or how products are marketed and distributed. There is no substitute for MRP, although future computer programs will be greatly simplified compared to early versions.

Simplicity comes not from major program changes but from streamlined manufacturing environments. The "rescheduling frenzy" of the past is unnecessary when orders flow swiftly through operations and when upsets and delays are few. Simplicity comes also from eliminating from MRP those items that can be controlled equally well and at lower cost and effort by means of physical controls.

Supplier-customer relations are changing radically and rapidly. Instead of the arm's-length, antagonistic (on the buyer's part), formal dealings, both parties are seeing the benefits of long-term partnership arrangements. Numbers of suppliers for each class of materials are drastically reduced; customers certify a single source for many categories, examining closely the supplier's operations and capability of close teamwork. Contracts are signed covering long periods, specifying aggregate demands on the supplier's capacity for families of similar items, and setting prices on individual items. Orders are released at the contract rate for individual items and delivered in small quantities in short lead times. MRP planners, often called *planner-buyers,* handle all communications with suppliers on order details. Both parties work constantly to improve relations and lower costs, sharing the benefits.

That management's commitment is needed for success is certain. There are three phases: first, develop its own understanding; second, change the environment to make the formal system effective; and third, use it to get tight control.

Developing management's understanding includes

1. Knowing how manufacturing should work
2. Learning the technology of planning and control
3. Seeing the fallacies in conventional wisdom
4. Recognizing the need for speed and flexibility

Changing the environment involves

1. Making partners of suppliers and customers
2. Identifying organizational groups' primary tasks
3. Streamlining organization and communications
4. Selecting correct measures of performance
5. Educating people at all levels

Using the system requires

1. Getting themselves in the feedback loop, and initiating corrective actions promptly
2. Reviewing performance of groups critically
3. Insisting on constant improvement

Detailed discussion of these topics relating directly to MRP and associated activities is included in this book. Topics not directly related are covered in references in the Bibliography.

Future Research Projects

Manufacturing presents many topics on which in-depth (i.e., not just more surveys) research is needed. Material requirements planning, the subject of this book, offers the following opportunities for productive research:

- I. Theory
    - A. Manufacturing lead time length and stability
    - B. Relations between safety stock and lead time
    - C. Links between planning and execution
    - D. Limits of MPS horizons and changes
    - E. Capacity requirements planning techniques
    - F. Input/output restrictions
    - G. Varying time-bucket lengths over planning horizons
- II. Program design
    - A. Design criteria for different business environments
    - B. Computer bills of material structuring
    - C. Simulations for problem solving
    - D. Simplification of MRP programs
    - E. Integration with data management programs and telecommunication networks

III. Implementation and use of MRP programs
   A. Costs of upsets in execution
   B. MPS development and management
   C. Planning system aids to execution
IV. Education
   A. Curricula content—primary and advanced
      1. Principles
      2. Techniques mechanics, strengths and weaknesses
      3. Planning and control versus execution
      4. Fundamental requirements—valid MPS, accurate data, qualified people, teamwork
      5. Performance measures—for MRP system and operations
   B. Problem solving with formal system data

Manufacturing lead time is discussed in Chapters 3, 4, 8, and 11 in detail, and referred to in other chapters. Chapter 11 stressed the compelling need to get and keep planned lead times short, although most users tend to want them long. Excessive planned lead times inflate the investment in work-in-process and diminish the effectiveness of priority planning and control. Education curricula should clarify the difference between planned (average) and actual (specific) lead times, and emphasize the value of making both shorter. Research for "optimum" values is a waste of time; *the best target is half the present values*. Research can identify the important elements in the value of shorter lead times; traditional cost systems do not yield these data.

Safety stock is useful for independent-demand items but unnecessary for dependent items planned by MRP. This topic is covered in Chapters 4, 6, and 11. Chapter 4 shows the logic of time-phased order points and Chapter 11 discusses their use in triggering fast reactions to forecast errors, making the accuracy of specific period data less important. The accuracy of the forecast average, however, is vital. Research could investigate new methods, replacing popular statistical ones, to produce better averages over long horizons. These should undoubtedly include better communication with customers. Safety stock can be reduced as lead times are compressed. The true relationship between these needs more investigation.

Links between MRP planning and execution subsystems are discussed in Chapter 10. These need to be defined better and the functions better integrated. Research can also study the interfaces between planning and execution, establish ranges in which execution can deviate from plans, and identify the possible penalties. Education curricula should make clear their different purposes and how the transition is made.

Planning is needed over long horizons but in aggregate totals, not details. *Projecting MPS over long future periods* is playing magic numbers games; they look good, but don't mean much. Study is needed to determine which factors are important in extending MPS beyond products' cumulative cycle times.

Capacity requirements planning techniques are either too crude (rough-cut) or too detailed (load profiles). Research into combinations of these two is needed to take advantage of their strengths and compensate for their weaknesses.

Input/output control monitors actual capacity against plans, and places restrictions on order release and rescheduling. Study of the factors and ranges of tolerable deviation are important.

The need for more specific data in the near term (hard execution data) and the inevitability of change over longer periods (soft planning data) make *different time-bucket lengths* eminently sensible. Research could be helpful in identifying ways to make transitions between different-sized buckets.

MRP programs are too complex and sophisticated. System designers have indulged their fancies in supplying "enhancements" to basic MRP. Simpler software is needed only with those options required to *adapt MRP logic to different businesses.* MRP program design features are presented in Chapter 3, but criteria for different applications have not been developed. Lack of rigorous criteria for user input data, including planning-horizon length, replanning frequency, time-bucket size, and lead times, and for including firm planned orders and time-phased allocations, results in these being selected empirically and intuitively. Research into optimum designs of MRP programs for various environments would be very pertinent.

Modular bills of material are discussed in Chapter 7, but *BoM restructuring techniques* to set them up are empirical. Principles and the mechanics could be formalized by research. Ideally, the logic of bill-of-material structuring can be described and programmed in computer software to do this task automatically. Even the ability to make only a first, rough cut would be valuable. Alternative BoM treatments of optional product features need to be explored and evaluated; these could aid both product design and master planning. Chapter 7 also covered "nested" options (option within option within option) and suboptions that require additional study to define limitations and benefits.

The use of *simulations for solving operations problems* has been a tempting topic for many years. Later in this section, experience with "hands-on" education using computer planning system data is presented and simulations are discussed. Much work is needed to identify where simulations can be helpful and how to apply them in dynamic operations.

Planning programs generally are too complex in well-managed businesses; *simplifying MRP* will improve their effectiveness. Modular software can include basic logic, common elements, and add-on programs as options for different companies. More resources and preproduction activities can be planned. Project planning can be improved by applying MRP logic; this is called Project Resource Requirements Planning. Little attention has been given to simplifying MRP; much more will yield great benefits.

Integrated data management programs provide data from massive databases when needed by other programs. These programs serve design engineering, processing, planning, and many other data users from a common databank, with great savings in storage, processing time, and cost of data handling. Errors can be reduced. The design and application of these valuable programs need attention from users as well as computer wizards.

Uses of MRP programs in operations, discussed in Chapters 9 and 10, have not been adequately investigated. What deficiencies does MRP have from users' points of view? What are the potential uses of MRP by inventory planners, capacity planners, master schedulers, and many others? What are the limits of information that MRP can provide? How timely should its data be? How can MRP best be used in making rigorous analyses of problems and helping develop solutions? What are the situations users commonly face, and what is the correct response to each one of these? Experienced MRP users have answered many of these and similar questions, but their answers are not recorded and made available to others. Their analysis, classification, and compilation would provide a useful guide to the use of MRP.

Problems causing upsets in manufacturing are legion. Their inevitability was the principal cause of MRP programs becoming over-complex and too sophisticated, attempting to cope with chaos by more powerful replanning. Backlash from the failure of this approach to work well damaged the credibility of MRP. Attention is needed to restoring this credibility by making MRP more effective. Minimizing upsets is the key, and carries high benefits from reduced operating costs.

Costs of upsets in operations are unquestionably high, but conventional accounting methods do not produce even good estimates. These include costs of unnecessary handling of excessive and behind-schedule material, poor or late operating management decisions due to lack of valid information, component staging, expediting by people with more important primary tasks, excess setups, and innumerable others. The true magnitude of these costs would indicate the potential benefits of making MRP programs work well. Field research into this would produce startling results.

Master production schedule development and management are discussed in Chapters 8 and 10. These represent potentially the richest "pay dirt" for research. Master scheduling has long been the weakest element in core systems. Management of this vital set of numbers has been left to people too low in the organization; top-level managers too often don't know MPS exist, and if they do, they don't understand their purpose or how to use them. Education curricula need to reach these executives and inform them of the "handle on their business" provided by MPS. Research is needed to clarify organizational relationships and interests, practical limitations on MPS changes, and the effects of overloading and underloading MPS.

Planners operating MRP programs frequently concentrate on planning actions with little regard to *aiding execution*. This can be done by providing capacity planning data over long horizons and order priorities over short periods, smoothing order input to match capacity, releasing a workable order sequence, and ensuring that all components, tooling, and other needed resources are available to work on released orders immediately.

Materials for education in this field abound. The literature in the body of knowledge of manufacturing planning and control is extensive, and that relating to MRP makes up a large share. The American Production and Inventory Control Society (APICS) publishes a monthly magazine and a quarterly journal, issues dictionaries and bibliographies updated periodically, and provides proceedings of all important conferences and workshops. Unfortunately, the quality of some of these materials is poor, even error-laden, with promotional and sales messages included in spite of formal prohibitions. Preparation of definitive materials is needed to provide authoritative bases for education.

Three phases of education must be provided:

1. The state of the art, encompassing the body of knowledge, language, principles, and techniques
2. Applications of these to specific solutions of problems in individual manufacturing companies
3. Specific skills needed by individuals working in the field in planning and control and related activities

The first can be handled by technical societies, consultants, trade associations, colleges and universities. Many now offer courses on operations, production, and inventory management, including MRP, using only texts written by the instructors, often of poor quality. There have been no well-defined standards governing materials, although syllabi for the APICS Certification Program have partially filled this need.

The second phase requires more specific instructors' knowledge of company needs. Well-trained company people, preferably supervisors and managers, are best; consultants with wide experience can do this well also.

The literature contains very little detailed discussion of *how data from modern MRP and related programs can be used to solve specific problems.* Course attendees at George Plossl Educational Services for years worked on minicomputers to learn how. Programmed with a simulated small-company manufacturing planning and control system, they used computer terminals to access data and solve specific data-related problems like

1. Make or buy this component?
2. A supplier will be two weeks late delivering a critical material; how will you minimize the effect on MPS?
3. When is the best time to phase in an engineering change?
4. A subassembly record has a significant error; how will you minimize the effects on its parents?
5. A bottleneck machine has broken down; what alternatives will ease problems until it has been repaired?

The average time to solve each of these problems was more than two hours! And these people had just finished an intensive course in principles, techniques, and how system elements interact. They were not phony problems; attendees admitted that they frequently had experienced problems just like these in their own companies. Reviewing their experiences, the consensus was that routine planning with programs like MRP was simple to learn, whereas problem solving was a completely different task calling for the answers to questions like

1. What data are needed to solve this problem?
2. What files have the data needed and how current are they?
3. What alternative solutions are available?
4. What items, people, and machines are affected, and how?
5. How can the system and people be told of changes needed?

Attendees agreed almost unanimously on two points: *computer simulation programs are unlikely to be of much help,* and *the only practical way to handle such problems is to eliminate as many as possible and get experienced people to solve the rest!*

The third phase is on-the-job training led by company people, usually the supervisors of those being trained. They need considerable training themselves on how best to develop the needed understanding, education materials, and teaching skills.

Creative research is needed both to sift wheat from chaff and in curricula redesign, development of teaching materials, classroom examples and exercises, case studies, and the construction of computer-based simulators for students. The APICS Certification Program for both members and nonmembers has made a good beginning, but much more attention is needed to promote education.

Conclusion

MRP has been a working tool of manufacturing planning and control for almost three decades. During this time it has been a mixed blessing. Some users have had outstanding results, a few have suffered disasters, and many have found MRP "not good enough to use but not bad enough to throw away." In spite of such spotty performance, many profound changes have taken place:

1. Early, widespread application of computers ended information-processing constraints on sound planning and control.
2. Development of time-phased material requirements planning, both a concept and a powerful tool, replaced crude, ineffective methods of inventory planning.
3. Knowledge of planning and control grew enormously.
4. Formal planning and control systems were developed that can function effectively in manufacturing without the props of informal expediting systems.
5. Truly professional approaches replaced traditional methods of "brute force and ignorance."
6. The loop between end-item master production schedules and component schedules was closed.
7. Misperceptions of manufacturing logistics were clarified.
8. Management awareness of the importance and feasibility of sound production and inventory control increased greatly.

Computer wizards and other idealists visualized for a long time a virtually perfect, detailed plan of material supply and production for any company. MRP brought such visions close to reality, but suffered from poor results. Idealists blamed people wanting changes, believing all would be well if only everyone would hold still until the plan could be executed.

The real world failed to cooperate with such dreamers. Even before planning had been completed, management changed master schedules, engineers changed product designs, sales changed forecasts, industrial engineers changed manufacturing processes, and customers changed their minds on what they wanted. Worse, when work actually started, Murphy's Law governed operations; what could go wrong did go wrong, at the least opportune time.

No one could stop the changes but dreamers still hoped, thinking that the problem was not poor planning but lack of good replanning. MRP supplied this capability also, and many believed that all would be well if only it were used instead of hot lists. Net change supplemented regeneration and some tried real-time, on-line net change. Replanning at the rate changes were taking place was impossible, however, and no planning tool, even one as powerful as MRP, could cope with chaos in manufacturing. Plans were still invalid, and people could not be held responsible for their execution.

Past experience proves that wishing for stability or depending on replanning are both false perceptions. Manufacturing is inherently unstable and turbulent. Change is "the name of the game" and will increase. Success lies neither in stabilizing and freezing plans nor in replanning at blinding speeds but in *eliminating unnecessary changes* (caused by lack of communications and problems that can be solved), thus enhancing the firm's ability to accept desirable changes, respond promptly and correctly, and do it routinely. Almost all plant production and data handling problems can and must be solved. To the surprise of many people, "surprises" by customers changing orders can be greatly reduced through better communications.

Other fallacies caused planning failures:

1. *Systems solve existing problems.* They don't; they just add new ones.
2. *Expensive systems "buy" the way out of trouble.* They won't; working out of trouble is the only way.
3. *Systems reduce the need for skilled people.* They don't; education in new understandings and skills is vital.
4. *Sophisticated systems will meet all needs.* They probably won't; good specifications must be developed first.
5. *Just do what the system says.* And fail at blinding speed. Human judgment must adapt plans to reality.

It is now clear what steps must be taken to implement an effective planning and control system:

1. Define the core and subsystems needed to meet the needs of the business.
2. Identify the elements critical to tight control, and make sure these are implemented soon and well.

3. Design the database needed by the system, and purge errors from existing data.
4. Link core and subsystems in an electronic network.
5. Set up performance measures with tight tolerances for all key activities.

The future will bring further enhancements in companies' ability to respond properly to change. These will involve expanded use of computer-based programs networked into integrated systems linking suppliers, manufacturers, distributors, and customers. Such networks already are being used successfully by major retailers; similar ones will help manufacturing firms enormously.

Teams of people from several functions, not individual all-stars, will attack problems and develop improvements continuously. All must be educated to the compelling need for elimination of waste of all kinds, speeding up and smoothing out flows of materials and information, and introducing greater flexibility to handle change. Formal systems can displace informal and be vastly more efficient in helping to run manufacturing the way it should be run.

Bibliography

This is a different type of bibliography. Rather than listing materials that cover the same topics as the book chapters, this lists books and articles that *supplement* the book's coverage, amplifying topics covered briefly, illustrating applications of MRP and its related techniques, and providing details of activities important to successful use of MRP but beyond the scope of this book.

Arnold, A., "Enterprise Master Scheduling," *APICS-LA/ADSIG 1991 Conference Proceedings,* Falls Church, VA.

Blocher, J. D., C. W. Lackey, and V. A., Mabert, "From JIT Purchasing to Supplier Partnerships at Xerox," *Target,* (AME), Wheeling, IL, May/June 1993.

Blood, Barry E. and Brian L. Blood, "How to Improve the Usability of Your MRP Output," *APICS 1990 Conference Proceedings,* Falls Church, VA.

Brown, Robert G., *Statistical Forecasting for Inventory Control,* McGraw-Hill, New York, 1959.

Brucker, H. D., G. A. Flowers, and R. D. Peck, "MRP Shop-floor Control in a Job Shop," *Production & Inventory Management* (Journal of APICS), Falls Church, VA, 2d Q, 1992.

Crow, Kenneth, "Advanced MRP Capabilities and Approaches for Defense Contracting," *APICS 1990 Conference Proceedings,* Falls Church, VA.

Crowell, J. K. and R. G. Ernst, "Engineering Activities Planning Using MRP," *APICS-AD/SIG 1991 Conference Proceedings,* Falls Church, VA.

Dickie, H. F., "ABC Inventory Analysis Shoots for Dollars," *Factory Management and Maintenance,* July 1951.

Fuller, T. H. Jr., *Microcomputers in Production and Inventory Management,* Dow Jones-Irwin/APICS Series, Homewood, IL, 1987.

Gue, F., *Increased Profits through Better Control of Work in Process,* Prentice-Hall, Reston, VA, 1980.

Guess, V. C., *Engineering: The Missing Link in MRP,* Vanard Lithographers, San Diego, CA, 1979.

Hall, R. W., *Attaining Manufacturing Excellence,* Dow Jones-Irwin/APICS Series, Homewood, IL, 1987.

Harty, J. D., G. W. Plossl, and O. W. Wight, *Management of Lot-size Inventories,* APICS Special Report, Falls Church, VA, 1963.

Klapper, N. D., *MRPII Software Evaluation and Selection, APICS: The Performance Advantage,* Falls Church, VA, February 1993.

Lankford, R. L., "Job Shop Scheduling," *APICS 1982 Conference Proceedings,* Falls Church, VA.

Lankford, R. L., "Victims of the Standard System," George Plossl Educational Services News Note 60, Atlanta, GA, 1986.

Ling, R. C., "The Production Planning Process—Top Management's Role," *APICS 1982 Conference Proceedings,* Falls Church, VA.

Lunn, T. and S. A. Neff, *MRP,* Business One Irwin, Homewood, IL, 1992.

Magee, J. F., *Production Planning and Inventory Control,* McGraw-Hill, New York, 1958.

Mather, H., *Bills of Materials, Recipes & Formulations,* Wright Publishing Co., Atlanta, GA, 1982.

Mather, H., *Competitive Manufacturing,* Prentice-Hall, Englewood Cliffs, NJ, 1988.

Mather, H., "Reschedule the Reschedules You Just Rescheduled," *Production & Inventory Management* (Journal of APICS), Falls Church, VA, 1st Q, 1977.

Orlicky, J. A., "Rescheduling with Tomorrow's MRP System," *Production & Inventory Management* (Journal of APICS), Falls Church, VA, 1977.

Orlicky, J. A., G. W. Plossl, and O. W. Wight, "Structuring the Bill of Material for MRP," *Production & Inventory Management* (Journal of APICS), Falls Church, VA, 4th Q, 1972.

Orlicky, J. A., *The Successful Computer System,* McGraw-Hill, New York, 1969.

Pelfrey, M. W., "Integration of MRPII, TQM, and CIM," *APICS-LA/ADSIG 1991 Conference Proceedings,* Falls Church, VA.

Plenert, G. J., "Bill of Energy," *APICS 1982 Conference Proceedings,* Falls Church, VA.

Plossl, G. W., "How Much Inventory is Enough?", *Production and Inventory Management* (Journal of APICS), Falls Church, VA, 2d Q, 1971.

Plossl, G. W., *Managing in The New World of Manufacturing,* Prentice-Hall, Englewood Cliffs, NJ, 1991.

Plossl, G. W., *Production and Inventory Control: Applications,* George Plossl Educational Services, Atlanta, GA, 1983.

Plossl, G. W., *Production and Inventory Control: Principles and Techniques,* 2d ed., Prentice-Hall, Englewood Cliffs, NJ, 1985.

Plossl, G. W. and O. W. Wight, *Material Requirements Planning by Computer,* APICS Special Report, Falls Church, VA, 1971.

Plossl, K. R., *Engineering for the Control of Manufacturing,* Prentice-Hall, Englewood Cliffs, NJ, 1987.

Plossl, K. R., "JIT, MRPII, and Other Fad Diets," George Plossl Educational Services News Note 56, Atlanta, GA, 1985.

Plossl, K. R., "Standard MRPII Software Derails JIT," George Plossl Educational Services News Note 57, Atlanta, GA, 1985.

Plossl, K. R. and E. Heard, "MRPII Fails to Yield High Inventory Turns," George Plossl Educational Services News Note 50, Atlanta, GA, 1983.

Sandras, W. A., Jr., *Just-In-Time: Making It Happen,* Oliver Wight Limited Publications, Essex Junction, VT, 1989.

Schonberger, R. J., *Japanese Manufacturing Techniques,* The Free Press (Macmillan), New York, 1982.

Sharma, K., "Adding 'Intelligence' to MRP Systems," from *APICS: The Performance Advantage,* Falls Church, VA, March 1993.

Shigeo Shingo, *Study of Toyota Production System,* Japan Management Association, Tokyo, Japan, 1981.

Sullivan, R., "Build a Better Planning Cycle," *APICS: The Performance Advantage,* Falls Church, VA, April 1993.

Suri, A., "Change Brought About by Time: What's the Impact on Planners and Schedulers?" *APICS: The Performance Advantage,* Falls Church, VA, January 1993.

Suri, A., "Master Production Schedule: The Driver of Planning and Control Systems," *APICS: The Performance Advantage,* Falls Church, VA, July 1992.

Theisen, E. C., Jr., "New Game in Town—The MRP Lot-size," *Production and Inventory Management* (Journal of APICS), Falls Church, VA, 2nd Q, 1974.

Wight, O. W., "Input/Output Control, A Real Handle on Lead Time," *Production & Inventory Management* (Journal of APICS), Falls Church, VA, 3rd Q, 1970.

Wight, O. W., *Production and Inventory Management in the Computer Age,* Cahners Publishing Co., Boston, MA, 1974.

Wight, O. W., *MRPII: Unlocking America's Productivity Potential,* CBI Publishing Co., Boston, MA, 1981.

Wilson, R. H., "A Scientific Routine for Inventory Control," *Harvard Business Review,* 13, #1, 1934.

Word, C. E., "Planning and Controlling Production-related Activities Using MRPII," *APICS-AD/SIG 1991 Conference Proceedings,* Falls Church, VA.

APICS Certification (CPIM) Program Study Guides (Instructor Guides and Student Guides), Falls Church, VA, 1991: Master Planning; Inventory Management; Material and Capacity Requirements Planning; Production Activity Control; Systems and Technologies; Just-In-Time.

APICS Certification (CIRM) Program Study Guides (Instructor Guides and Student Guides), Falls Church, VA, 1992: Customers and Products; Logistics; Manufacturing Processes; Support Functions.

Index

About the Author

George W. Plossl (Fort Myers, Florida) is one of the pioneers of manufacturing planning and control. A highly respected manager, consultant, and lecturer, Plossl has authored several leading books in his field. During his illustrious career spanning four decades, he has worked with nearly every company in the Fortune 500 and in most of the world's industrialized countries. He has addressed almost every American Production and Inventory Control Society chapter in the United States and most of its foreign affiliates. Listed among the Who's Who in Industry and Finance, he is the only pioneer in the manufacturing control and planning field still active. He was elected one of the first honorary members of APICS.